S0-BBR-423

Pushing Our Limits

MARK NELSON

Pushing Our Limits

Insights from Biosphere 2

THE UNIVERSITY OF
ARIZONA PRESS

TUCSON

The University of Arizona Press
www.uapress.arizona.edu

© 2018 by Mark Nelson
All rights reserved. Published 2018

ISBN-13: 978-0-8165-3732-7 (paper)

Cover design by Carrie House
Cover photograph: *Biosphere 2 at Night* by Gill C. Kenny

Library of Congress Cataloging-in-Publication Data
Names: Nelson, Mark, 1947– author.
Title: Pushing our limits : insights from Biosphere 2 / Mark Nelson.
Description: Tucson : The University of Arizona Press, 2018. | Includes bibliographical
    references and index.
Identifiers: LCCN 2017042851 | ISBN 9780816537327 (pbk. : alk. paper)
Subjects: LCSH: Biosphere 2 (Project) | Ecology—Research.
Classification: LCC QH541.27 .N45 2018 | DDC 577.072—dc23 LC record available at
    https://lccn.loc.gov/2017042851

Printed in the United States of America
♾ This paper meets the requirements of ANSI/NISO Z39.48-1992 (Permanence of Paper).

Contents

Preface

What Have I Gotten Myself Into?

ON A WINTER NIGHT in January 1993, by opening a doorway we experienced a stunning physiological revival. We left a world with an oxygen level around 14 percent; equivalent to being on a 15,000-foot-tall mountain. In fact, we were at a 3,900-foot elevation in southern Arizona. Oxygen had been slowly disappearing for sixteen months. No one knew where it had gone. We were slowly climbing a mountain but going nowhere. Mission Control had pumped oxygen into a chamber on the other side of the door. Our atmosphere suddenly contained 26 percent, which was 5 percent higher than Earth's air. In minutes, we felt decades younger. For the first time in many months, I heard the sound of running feet.

So many strange, disturbing, marvelous, powerful, and profound experiences unfolded during our two years as "biospherians." The eight of us felt extraordinarily lucky to be the initial crew to live inside a miniature biosphere. We had to learn how to be its first natives.

Biosphere 1 (B1) is our Earth's biosphere. Biosphere 2 was a three-acre world. B1 houses the global ecosystem, which includes all life. B1 is our planet's life support system. Biosphere 2 was built to study how biospheres work, creating a laboratory for global ecological processes, to help ecology become an experimental science. It could also provide baseline information to design long-term life support systems for space.

The facility included people, farming, and technology. Earth's biosphere has supported life for four billion years. Only quite recently have billions of people and modern industries been added. Living in Biosphere 2 might give new perspectives on whether—and how—harmony can be forged between humans and the global biosphere.

Our two-year experiment began on September 26, 1991. We'd have two seasonal cycles to study how Biosphere 2 functioned. For comparison, a human spaceflight to explore Mars would also take two years. No one knew if we could stay inside for two years; so many things could go wrong. The facility was optimistically designed for a one-hundred-year operation. The first closure experiment was the "shake-down" mission; a trial run to find flaws, bugs, what we had to correct or change. We were also determined to collect as much data and to do as much research in collaboration with outside scientists as possible.

The odds, even from project insiders, heavily favored an early exit. Too many challenges—known and unknown—could end the experiment early. Some thought we wouldn't even last three months. The world record in a closed ecological system was six months set by two-person crews at a Siberian research institute. Their basement facility powered by artificial lights was the size of a small apartment, their only companions were food crops. Our own sunshiny world contained a rainforest and a coral reef in a towering structure with seventy-five-foot-tall roofs. Every day we were able to stay alive inside, we would amass reams of research data.

We entered an untested facility in almost totally uncharted territory.

We included small chunks of Earth's diversity inside the biosphere; bonsai rainforest, tropical grassland (savanna), desert, mangrove marsh, and coral reef ocean co-existed under one roof. Some of the world's top ecologists and most innovative engineers worked to make this possible; no one knew how these biomes would develop. Ours was cutting-edge science, the greatest experiment in ecological self-organization ever conducted. To maintain biodiversity, we biospherians would intervene when we could. Our fog desert decided to go its own way and transformed during the experiment; maintaining the others took hard work and ingenuity, the coral reef in particular, was a nail-biter to the end.

In our nearly airtight world, we would experience the highs and lows of living intimately with seven other people. Outside politics and power struggles polarized and exacerbated in-fighting, though we entered as the best of friends and colleagues. I wouldn't permit a bitter "To the traitors" as toastmaster at a Sunday night dinner where we enjoyed a precious bottle of home-brewed banana wine. There were no fistfights, but one crew member complained years later that she had been spit at. *Twice*. But we continued working unselfishly with one another. Whenever we feasted, partied, or enjoyed a rare delicacy like a cup of coffee from rainforest trees, tensions magically melted away. We'd relax and enjoy a temporary truce from group tensions. We acted mindfully in Biosphere 2, understanding that its teeming life was keeping us alive and healthy. We took care of her needs with tender loving care. She was our third lung and lifeboat. Some of us thought Biosphere 2 was the ninth biospherian.

Eight Americans and Europeans suddenly became subsistence farmers. We lived off the land, eating what we grew, though we farmed in a high-tech, $150 million facility. Our small farm exceeded organic standards. We used nothing that might pollute our air, water, soils, or crops. We recycled our water and soil nutrients. Even our sewage was treated and recycled. We cared for our farm animals with affection, but they were slaughtered as needed. Our diet consisted primarily of fruits, grains, and vegetables.

We experienced hunger throughout the two years and plates were always licked clean. Almost all of us became much better cooks. Peer pressure for delicious food was a great motivator. I and many others ate our roasted peanuts whole, shell and all; we would eat anything to fill the stomach void. We were guinea pigs, the first humans extensively studied on an "undernourished but not malnourished" diet. This paralleled the pioneering research of Biosphere 2's in-house doctor, who claimed a person could live 120 years on a calorie-restricted diet.

Periodically, project managers reminded us we were volunteers; the airlocks were unlocked, and we could leave anytime we'd had enough or if there were health dangers.

For safety, we had our resident doctor and a team of specialists on call at the nearby University of Arizona College of Medicine, and a fully equipped medical facility and analytic laboratory were inside the biosphere. Automated systems could detect potentially toxic substances in our air and water. We started with a biosphere as clean and unpolluted as possible. Chemical deodorants and cleaning supplies weren't allowed because our world was so sensitive to pollution. Even a small fire would mean evacuating, so we didn't light candles, even on a birthday cake. At winter parties, a monitor played a video of a wood-burning fireplace—we felt warmer sitting near it.

Though we didn't intend it, the toes of dominant analytic, small-scale science were seriously stepped on. The reductionist approach seeks to analyze everything at the micro level, each variable being tested separately. Biosphere 2 used both analytic and holistic science approaches. The project violated unspoken taboos. Include humans and our technologies *in* the experiment? Heresy! We knew one thing for certain: Biosphere 2 would ignite plenty of controversy.

Systems ecologists and veterans of NASA's Apollo Project 1960s glory days were allies from the beginning. To achieve the goal of putting a man on the moon by the end of the 1960s, NASA abandoned component-by-component testing and went to "all-up systems testing." We followed a similar strategy to create this complex miniworld; it couldn't be done piece by piece like Lego.

The six years from project conception until completion were exciting. Scientists, engineers, and hundreds of construction workers were very motivated. They were making history, doing the near impossible. Some doubted at every stage whether Biosphere 2 could be built, operated, or used for advancing human knowledge. Who were these mavericks behind the project? Despite many world-class scientists and institutions consulting, the whole endeavor was way too ambitious, too daring. Even some friends and colleagues of the project thought it was fifty years ahead of its time.

Biosphere 2 was radical and revolutionary—a challenge to "business as usual." The entire "technosphere" had one overarching goal: serve and protect life. Our engineers had to design technology to make waves, rain, winds; they had to control climate and mimic geological processes. And they had to use machinery and equipment that wouldn't poison and pollute. Life ruled. Technology knew its place and obeyed and served, a radical notion. What would happen if we did that everywhere?

The engineering goal was about 1 percent per month air exchange (leakage) from the biosphere. That's thousands of times tighter than the most tightly sealed buildings and homes, far tighter than even the International Space Station. But, if this air-tightness was achieved we might wind up with a horrific "sick building syndrome" from a buildup of trace gases. We needed a way to ensure that those trace gases didn't build up in a structure with two acres of farm and wilderness areas, hundreds of pumps, motors, and other equipment, and miles of piping. Our solution was to use our farm soil and plants as a biofilter to clean the air. We hoped it would work.

Carbon dioxide was called the tiger of Biosphere 2. We continually monitored its levels in our atmosphere since it could destroy our world, and it would be difficult to keep the levels from rising too high. Every cycle goes hundreds to thousands of times faster than normal in a tightly sealed, small, and life-packed miniature biosphere. Our ocean and atmosphere were tiny compared to Earth's; we had entered a time machine.

Would all the life inside Biosphere 2—with us humans doing everything we could to help—be sufficient to prevent a runaway rise in carbon dioxide, our tiny version of climate change? If CO_2 levels got too high, our coral reef might die, all the plants (including our food crops) might slow their growth, and our health might be directly threatened.

By closing the airlock behind us and starting our two-year experiment, we pushed the limits and stepped into the unknown. It would be a roller coaster, with despair and sadness and euphoria and achievement. Every day, we worked to keep Biosphere 2—and ourselves—alive and healthy. For the eight of us, it was a profoundly personal and life-changing journey.

Biosphere 2 Revisited: What Lessons Does It Have for the World?

I exited Biosphere 2 more than twenty years ago.

Humans have caused the ongoing planetary crisis. Can we learn to live sustainably and happily within our own biosphere? Lessons and insights can be gained from a fresh look at how our small team handled our miniature biosphere. Our life and well-being depended on maintaining a healthy world.

Biosphere 2, in my opinion, is widely known but little understood. Close to one billion people around the world followed our experiment as it unfolded. The story has mostly been superficially and incompletely told.

In retrospect, some headlines that were noteworthy at the time now seem insignificant: Jane Poynter severed a fingertip in a threshing machine accident three weeks into our experiment. She brought with her a duffel bag of computer parts when she returned from surgery, not food as some alleged. Whether we grew 100 percent of our food, if everything was perfect and balanced on the first try are superficial ways of evaluating Biosphere 2.

Biosphere 2 was an experiment. An experiment fails if it doesn't reveal something unknown and unexpected. Needing to inject oxygen sixteen months into the closure was one of the most surprising things that happened. Losing atmospheric oxygen so slowly showed we achieved near total airtight sealing. We learned so much solving the mystery of where the oxygen had disappeared. Despite Biosphere 2's complexity, using the tools of science, its dynamics could be deeply investigated. Another surprise was how well the engineering and ecological design meshed together. But these topics don't make for sensational or viral news stories. The real news was that what was learned in Biosphere 2 has increased its importance to our global ecological challenges.

The Overview and Innerview Effects

Pondering our relationship to Biosphere 1 led me to recall how space exploration changed human consciousness. Photos of Earth from the darkness of space stunned the world. For the first time, humanity saw the planet, this spinning blue and white world that contains everything we value. Apollo astronauts enjoyed "Earthrise" over the horizon of the moon. Space voyagers' unique experience was called the Overview Effect.[1] It was a thrilling, life-changing perspective; they were the first humans separated from Earth's biosphere.

Rusty Schweickart flew on Apollo 9, testing the Lunar Lander. He served on the Biosphere 2 project review committee and became a friend.

Rusty eloquently described standing on top of a spacecraft orbiting the Earth:

You look down there and you can't imagine how many borders and boundaries you crossed again and again and again. At that wake-up scene—the Middle

Figure 1. Conversing with Rusty Schweickart at a Biosphere 2 design workshop, 1987.

East—you know there are hundreds of people killing each other over some imaginary line that you can't see. From where you see it, the thing is a whole, and it's so beautiful. And you wish you could take one from each side in hand and say, "Look at it from this perspective. Look at that. What's important?" . . . And you realize . . . that you've changed. That relationship is no longer what it was . . . Do you deserve this? This fantastic experience? You know very well and it comes through to you so powerfully, that you're the sensing element for man . . . It's a feeling that says you have a responsibility. It's not for yourself . . . And when you come back, there's a difference in that world now, there's a difference in that relationship between you and that planet, and you and all those other forms of life on that planet, because you've had that kind of experience. It's a difference, and it's so precious . . . it's not me, it's you, it's us, it's we, it's life that had that experience. And it's not just my problem to integrate, it's not my challenge to integrate, my joy to integrate—it's yours, it's everybody's.[2]

Those images and words from the first voyagers beyond Earth were revolutionary. Now, they're part of the fabric of reality for younger generations.

The Overview Effect, which the astronauts shared, was the crucial beginning of a realization that the whole Earth is our home.

But biospheres are still a different matter. The joint venture company behind Biosphere 2 was Space Biospheres Ventures, and I usually had to spell the middle word to people. Times have changed, and biosphere is used more often. I wonder though what people think when they say "Earth's biosphere." It is a hard concept to comprehend. Our biosphere is extremely complex; its cycles of air, water, and food enable human life along with all other life. We humans are an integral part of it. The biosphere is not some alien outside environment. These realities mean the current assault on and degradation of our biosphere pose almost unimaginable risks. Astronauts who deliberately damage their life support system would be considered suicidal nuts.

The fifteen who lived inside Biosphere 2 in missions one (1991) and two (1994) and others who lived inside even smaller closed ecological systems had the kind of life-altering experience that spacefarers report. But ours is not an Overview Effect, but an "Innerview." We lived with, took care of, and were nurtured by a living system—a miniature biosphere. All the while, we received communications and listened to news of developments between the rest of humanity and the planetary biosphere. As our connection with our world deepened, we became acutely aware of the jarring contrast between how we lived inside the biosphere and what was happening outside it. We'd hear of ecological degradation, mindless pursuit of technological advance, and short-term profit, regardless of the damage inflicted. We inwardly echoed what Schweickart wanted to say to people fighting over national boundaries: "Look at our biosphere from this perspective. What's truly important?"

I want to share this amazing experience that my colleagues and I were so fortunate to have—on behalf of you, the reader, and all life. I'll tell this story from a very personal viewpoint because I was one of eight humans, eight sensing elements in a new relationship to a biosphere, experiencing the Innerview Effect. The joys, struggles, and insights of that inner journey need to be communicated and integrated with one very large challenge: learning to take care of our global biosphere.

Acknowledgments

TO MY BIOSPHERIAN crew mates, Abigail (Gaie) Alling, Linda Leigh, Taber MacCallum, Jane Poynter, Sally Silverstone, Mark (Laser) Van Thillo, and the late Roy Walford: I am proud to have shared this journey with you. This book is a celebration of what we accomplished.

To John Allen, for dreaming no small dream, for having the vision and the can-do to make it real.

To Edward Bass, for generously funding Biosphere 2 through its several eras and his continuing dedication to the environment.

To Margaret Augustine, for unflinchingly guiding the project from inception to success.

To my colleagues at the Institute of Ecotechnics, for decades of passion, commitment, and ingenuity. For taking on the hard problems and having fun on the way. "Honor, beauty, discipline, and friendship."

This book centers on the eight of us who lived in Biosphere 2. But we could not have succeeded without the help of all those in mission control, the project's leadership and management, and dozens of scientists and engineers. They all contributed their expertise and innovative thinking to solve the problems that arose and enriched our research program.

And to the University of Arizona, for giving Biosphere 2 a new life and ensuring its continuing role in public education and cutting-edge science.

Pushing Our Limits

1

Build a Biosphere

What Were We Thinking?

WHAT STARTED US thinking about new worlds? Why did we decide that actually building one made sense?

Let's start with the simplest—possibly the most fun—explanation. A rumor circulated during construction of Biosphere 2. Oracle, Arizona, the nearest town, is a small, sleepy community that circulated the conspiracy theory that building a biosphere was just a cover story. The real reason for the biosphere? We were extraterrestrials and wanted to return to our galaxy. On closure day, Biosphere 2 and its homesick ETs planned to take off into deep space. Great story! But the true story is a more complex one.

Comparative Biospherics

Dr. Jack McCauley, a United States Geological Survey astrogeologist, spoke in 1978 at the Institute of Ecotechnics' (IE) annual conference. His talk rocked worldviews. He showed photos of Earth deserts and deserts on Mars. Wind erosion mainly, but also water erosion, sculpted beautiful land-scape forms on both worlds. We couldn't guess which planet was which. He declared we're living in the greatest age of exploration; humanity was learning so much about this amazing solar system, galaxy, and universe we inhabit. Columbus, Magellan, Cook—they were pikers by comparison. Maps of Mars were more detailed than maps of some regions of Earth!

A whole new field was opening: comparative planetology.[1] The space race propelled the United States and the Soviet Union to outdo each other with spectacular firsts: explore the rings of Saturn, land on Mars and Venus, orbit the Sun, journey to the ends of the solar system and beyond. Spacecraft discovered oceans of water and more exotic chemicals on other planets' moons. Humans saw images of volcanoes erupting on a world no eyes had seen before.

McCauley described the awe he felt, sitting in an auditorium at the Jet Propulsion Laboratory in Pasadena, California. Line by line, the first images of distant worlds sent back by spacecraft were revealed to the assembled space scientists. The exploration of Earth from space revolutionized our understanding of our planet and biosphere, and this mission now deploys far more, and increasingly sophisticated, orbiting sensors. Space exploration powered a global communications revolution, making the Internet possible. Comparing planets and learning how Earth differs from our neighbors is science fiction made real. As members of the Institute of Ecotechnics, we began to think of the possibility and promise of comparative biospheres.

We have only one biosphere to study until we find others (or they find us!) among the planets that orbit most stars. We started calling Earth's biosphere by the name "Biosphere 1" to emphasize how incredibly precious it is. Miniature biospheres could open new methods of investigating the basic laws and mechanisms governing all such systems. Though inherently complex, minibiospheres could be studied in far greater detail than is possible with our planetary system. All biospheres operate with photosynthesis and respiration, the water cycle, biogeochemical cycles, food chains and species interactions, soils and aquatic systems, and microbial diversity, and they respond to key environmental parameters like light, temperature, and seasonal changes.

Minibiospheres can be studied at many levels—species, populations, ecosystems, biomes—to see how ecological adaptation and self-organization unfold. They could function as laboratories to investigate the impact humans and our farming and technologies have or how one biome affects another and its impact on the overall biospheric system. The macroscope and the microscope: they could be holistic *and* analytical scientific instruments. Experiments involving dangerous chemicals or "destructive testing" extreme events, like drought or elevated temperatures, could be carried out in minibiospheres that we wouldn't and shouldn't do with our *one* global biosphere.

Space Life Support

We saw opportunity given the status of closed ecological systems and space life support using biology rather than machines. Early research tried to achieve complete life support with the single species chlorella algae. Both NASA and the Soviet space program failed. The algae were good at air and water regeneration, but people got sick eating more than an ounce of green algae. In disappointment, NASA shut down its research program for more than a decade. But the Russians did the obvious, including normal crops, from lettuce to wheat to potatoes, in their closed ecological systems.[2]

The Institute of Biophysics in Krasnoyarsk, Siberia, developed the most advanced Russian system for space life support, Bios-3. Up to three people could live in the materially sealed structure, getting their food from a dozen food crops grown hydroponically under high-intensity lights. Sometimes they included algae tanks to help with air and water regeneration.

Bios-3 recycled air and water, but solid human wastes were exported and high-protein meat products were sent in. Eighty percent of their food was grown inside, and they were able to achieve six-month closures.[3] These scientists were years ahead of NASA then and even now.

On a solo journey in 1985 to Russia, I met academician Oleg Gazenko, director of the Institute of Biomedical Problems (IBMP), a leading space biology institute. We began to lay the groundwork for cooperation.

Vernadsky Opens the Door to Biospheric Theory

In London in the early 1980s, we obtained a translation of selections from the Russian scientist V. I. Vernadsky's pioneering book on the biosphere, then unavailable in English. I was carrying that translation when I first met academician Gazenko. I inquired if he or one of his colleagues would write an introduction to this first English publication of his work. Dr. Yevgeny Shepelev, the first man to live for twenty-four hours in a closed ecological system, did so. (Gazenko, who became a dear friend, cut a deal for the introduction to the book: "Bring me a bottle of excellent Scotch whiskey the next time you come to Moscow!")

A year later, with a delegation from the Institute of Ecotechnics and Biosphere 2, we met the Russian leaders in the fields of closed ecological systems and space life support. Some at NASA headquarters doubted we would ever meet the scientists behind the Bios-3 project because of its secrecy. We even heard it might be Soviet propaganda and didn't actually exist. But during our meeting, Gazenko introduced a broadly smiling surprise guest, Dr. Josef Gitelson, head of the Institute of Biophysics (IBP), Krasnoyarsk. He carried a large film canister and opened our days of discussions by showing us a Russian television documentary about their work in Siberia.

The Russian scientists, both at Gazenko's IBMP and Gitelson's IBP, immediately saw the importance of what we were proposing. Vernadsky had elu-

Figure 2. The Russian and American biospherics and closed systems pioneers meet. Foreground: Oleg Gazenko, head of IBMP in Moscow, and I discuss collaboration with the most advanced Soviet space research institutes. At end of table: John Allen speaks with Yevgeny Shepelev, the first man to live inside a closed ecological system.

cidated our modern understanding of the biosphere in his 1926 book, *The Biosphere*.[4] He viewed the biosphere as not just a lucky passenger on Earth, but the powerful force that has transformed the planet. This insight led Vernadsky to found the science of biogeochemistry.

Vernadsky's ideas preceded the Gaia hypothesis developed by James Lovelock and Lynn Margulis about the role of life in maintaining a livable planet.[5] Founder of more than a dozen institutes and revered as a visionary and moral exemplar, Vernadsky's work influenced generations of scientists in Russia and was the key that opened up our project's alliance with Russia; they were excited that Biosphere 2 would move closed ecological system science to a truly biospheric level.

From Vernadsky's successors came important insights for the ecological design of Biosphere 2. Kamshilov, in *Evolution of the Biosphere*, showed that biomes were the building blocks of the biosphere.[6] Biomes are large regional assemblages of plants and animals characterized by a particular climate and dominant vegetation: rainforest, desert, tundra, grassland, coniferous forests, etc. Biomes interact and shift from one state to another if climatic conditions change. Previous life support systems had only included two human-dominated biomes: urban, where people live and work, and agriculture, where food is grown. No one had ever tried to make a biosphere before. To make a representative biosphere that could function as a simplified model of our planetary biosphere, we would need to include a range of systems modeled on natural biomes.

In 1989, a group from Biosphere 2 and other space and ecological scientists would be the first outsiders to visit the Siberian Bios-3 facilities. Gitelson's IBP hosted the Second International Conference on Closed Ecological Systems and Biospherics, which IE helped organize. The first conference was held in 1987 at the Royal Society in London. Our Russian colleagues were clever politicians; you had to be smart to do creative work under the endless red tape and restrictions of the Soviet system. When Mikhail Gorbachev, the head of the Russian government, visited their institute, they showed him telexes we had sent. They excitedly lied and said that they had just been received. They lobbied him, saying the outside world had great interest in their work. Gorbachev was promoting glasnost, making the Soviet Union more open. He agreed to allow international scientists entry. Krasnoyarsk, Siberia, had been a closed city even to Soviet citizens because its military

Figure 3. The First International Conference on Closed Ecological Systems and Biospherics, 1987, Royal Society, London. Roy Walford asks a question of Josef Gitelson (at podium), IBP. Seated from left: Margaret Augustine, CEO of Space Biospheres Ventures; Ganna Meleshko, head of the closed ecological systems laboratory at IBMP, Moscow; Clair Folsome, University of Hawaii; William Knott, head of the bioregenerative life support program at NASA Kennedy Space Center. Chairing the conference, I am standing on the right. (Photo by D. P. Snyder.)

infrastructure included radar that may have violated the Anti-Ballistic Missile Treaty.

We gained enormous expertise from cooperation with the IBP and IBMP scientists. Visiting scientists shared their studies and experience in closed ecological systems, accelerating our research and development programs.

Mesocosms and Laboratory Ecospheres

Prior work with open ecological microcosms and sealed ecospheres gave us confidence that a small biosphere could prove to be a useful scientific tool—and might also work! Closed ecological systems build on the long history of ecologists engineering microcosms and mesocosms to study natural systems. These small models of natural ecosystems have air exchange and replace water lost through evapotranspiration. But in these, scientists are able to control and manipulate major environmental vectors. Such model ecosystems have yielded important ecological insights.[7]

We hoped to reintroduce ecological approaches to the challenge of space life support. In the 1960s, NASA debated varying approaches to creating bioregenerative life support. By using biological processes, the scientists would be able to accomplish life support needs like growing food and cleaning and recycling water and air. Two influential ecologists, professors Eugene Odum (University of Georgia) and H. T. Odum (University of Florida) led a group urging an ecological approach.[8]

The ecologists lost out. Instead, NASA mainly relies on high-tech, super-controlled engineering approaches to space life support, though they now include food crops in their research program. H. T. Odum is considered the father of ecological engineering, using natural processes in a scientific, engineered way to solve environmental problems. The Odum brothers founded systems ecology, taking a holistic, multidisciplinary approach that studies ecosystems as complex systems with emergent, evolving properties. Both became great supporters of Biosphere 2. They saw it as an exciting extension of the ecological mesocosm approach as well as a unique test bed for ecological engineering.

Biosphere 2 also attracted the help of Dr. Clair Folsome, a microbiologist at the University of Hawaii. Starting in the early 1960s, he became the first scientist to enclose microbial ecosystems in sealed flasks. He, along with a few other scientists, discovered that the natural microbial diversity found in seawater or ponds is sufficient to keep such materially closed "ecospheres" cycling and living for decades. These closed ecospheres continue to live as long as they receive light and aren't overheated.[9]

Folsome and his team pioneered modern laboratory-sized closed ecological system research. Their earliest predecessor, Joseph Priestley, enclosed a mouse in a glass jar in the eighteenth century. It died. Then he added a plant, and both survived. Priestley invented the world's first closed ecological system and discovered oxygen, the mystery gas the plant released that allowed the mouse to live.[10]

Astronautics and Biospherics

Our ability to live in space was significantly lagging behind astronautics, which was rapidly expanding its capabilities for launching spacecraft and

Figure 4. Clair Folsome of the University of Hawaii with one of his sealed eco-spheres, stocked with a diverse aquatic ecology.

exploring space. NASA restarted its work in bioregenerative life support in the late 1970s with the CELSS (Controlled Environmental Life Support Systems) program, but this work has never been given priority status and is perennially underfunded. NASA and other space agencies opted for sending the essentials for life into space as consumables—frozen meals, tanks of oxygen and water—rather than producing and recycling them on board. Human waste is either discharged into space or brought back freeze-dried.[11]

The Russians did far more ground-based development, but their life support systems have never been tested in space. Some small test chambers in both the American and the Russian space programs have studied how individual plants and animals respond to space conditions including microgravity. Both biospherics and astronautics are required for long-term exploration and eventually habitation off the planet.

Another issue is whether simplified life support systems provide an adequate model for living in space, in orbiting space colonies or on Mars and the moon. Would a mechanized system for growing hydroponic crops provide adequate ecological resilience and meet human psychological needs? Our contention is that biospheres are the natural habitat of human life. We live in a biosphere, and eventually what will be needed are space biospheres if we are to live for extended periods off-planet.

I also think of biospheres in space as a payback to the biosphere that birthed and nurtured our species. Humans are the most technologically gifted species, and we will take Earth life with us to space. We have to. Lynn Margulis and Dorion Sagan wrote about the significance of Biosphere 2 in *Biospheres from Earth to Space*.[12] The creation of minibiospheres marks the first time our biosphere might give birth to a living propagule capable of reproduction.[13]

Joining the Space Race

John Allen, co-founder and director of the Institute of Ecotechnics, came up with the idea of Biosphere 2. He is rightly credited as its inventor. Our intrepid IE members had been unfazed by previous challenges to start cutting-edge projects around the world in ecologically difficult regions.

We thought such a visionary idea could attract the financial support and network of top scientists needed to carry out such a daring project. IE decided to join the space race!

As a preliminary exercise in biosphere design, John Allen assigned architect and IE collaborator Phil Hawes the challenge of presenting a spacecraft with internal biological life support at the 1982 Galactic Conference at our ecotechnics project and conference center in Aix-en-Provence, France. Creative groups of IE members helped by brainstorming how plants, animals, soils, and recycling systems could function as a life support system inside the craft for the presentation. Hawes, perhaps drawing on his background in New Mexico architecture, solved the radiation hazard by suggesting a three-foot layer of adobe around the interior. If NASA engineers had been at the workshop, they'd have been apoplectic considering the cost to launch every pound from Earth's gravity well.

But the creative work was fun and revealed some ecological sophistication. Jane Poynter, one of the biospherians, recalls Buckminster Fuller, the inventor of the geodesic dome and a hero who also spoke at the meeting, rose to comment, "I didn't think you could do it, but what you've proposed here does make sense." Addressing the members of IE, Fuller continued, "If you don't build the biosphere, who will?"[14]

We expected Biosphere 2 to be a quiet research facility. Southern Arizona can top 100 degrees Fahrenheit for months on end in the summer, so the choice of Oracle, Arizona, for the site was a poor one if the aim was to attract visitors. The renown that the project eventually gained, while unanticipated, perhaps reflected people's response to exciting real-time science and exploration of the unknown. Around the world, fascination grew about our lives inside our new biosphere. Once people, uninvited but galvanized by early press stories, started coming in droves, we implemented tours and a visitors' program. Their immense excitement and curiosity made us realize there's a pressing need for education in ecology and biospheres. The revenues could also help defray costs.

Biosphere 2 was a privately funded venture. Edward Bass, who had worked with IE for more than a decade, generously committed to financing the project. Bass has contributed to a number of worthwhile ecological conservation and education institutions and projects, both before Biosphere 2 and in the years since.

We hoped to recoup the investment by developing green technologies as well as creating other biospheric systems that could be located in major cities or tourist destinations. We envisioned every country and major university one day having its own biosphere for ecological education and basic research. One day there might be Biosphere 3, Biosphere 4 . . . Biosphere 101. Each biospheric facility could have a differing set of biomes, and each could experiment by manipulating starting configurations and environmental conditions. This would deepen our understanding of biospheric processes.

After the fall of the Soviet Union, academician Gazenko and other leaders of their ecological and space sciences hoped to build a Russian biosphere. But unlike Biosphere 2, it would start out quite polluted (as their country was at that time). They could focus their research on methods of cleaning soils, water, and air through bioremediation. Biospheres in northern climes could choose from temperate or polar biomes: tundra, boreal forest, and Arctic ocean, rather than the tropical and subtropical ones we chose for Biosphere 2 in the warm southwestern United States.

Developing educational curricula could also generate revenue. A good deal of our motivation was idealistic. Biosphere 2 was a privately funded project with public benefits as a goal: education, research, and technology development. We had not anticipated that it would reach and inspire hundreds of millions of people around the world. That was a tribute to an idea whose time had come.

Experimenting with Small Biospheres, Not Our Life Support System

In a sense, we humans are unconsciously and inadvertently conducting a vast experiment with our planetary biosphere.

How many people can it support and at what level of resource consumption? Given resource extraction and conversion of vast areas of the planet, what percentage of natural biomes can be altered for human use before problems arise? What happens when tens of thousands of new synthetic chemicals produced in vast quantities are introduced into basic air, water, food, and soil cycles? Biodiversity is projected to drastically diminish if there isn't a dramatic course correction. Is there a need for forests, coral reefs, grasslands, or any other living systems beyond us and our domestic animals and crops? Humans use an increasing share of the planet's fresh water and energy. With inadequate preparatory research, the impacts of genetically modifying organisms and releasing them are yet to be determined.

The atmosphere is a heavily human-impacted system as increasing greenhouse gases and acid rain from industrial pollution affect our world. This unplanned planetary "experiment" may include other unintended consequences of our unprecedentedly large and industrialized population. Some place their hopes on a "techno-fix" for whatever problems we cause. It's uncertain whether we will even have adequate time to respond.

There are serious moral issues about our current attitudes that impact the biosphere. Do other species and ecosystems have any inherent rights? Are there any limits we must respect when altering the biosphere? This global experiment has stakes that are incomparably high—our health and well-being, along with that of all other life on Earth. The planetary biosphere is quite literally our life support system, even if we act as if this is not the case.

So, the Biosphere 2 creators contended that small biospheric laboratories should be the place for experimenting with biospheres. Controlled environments can be used to see what GMO crops will do in nature or the pathways and impacts of a new chemical, pollutant, or technology. What we learn from closed ecological system chambers and biospheric laboratories can give insights into the consequences of our actions in the global environment and lead to changes.

Harold Morowitz, the distinguished biophysicist from George Mason University, noted: "Biosphere 2 provides, for the first time, the possibility of conducting controlled, large-scale ecosystem ecology experiments. Modern Physics emerged when Galileo conducted experiments that yielded numerical data. Biosphere 2 provides a setting for the same type of transformation in ecology as occurred in physics."[15]

In this context, the $200 million and the decade of work it took to create and launch Biosphere 2 are a bargain (and comparable to the cost of one fighter jet). Given the threats to our biosphere, Biosphere 2 is an assertion that it's time for ecologists to have experimental facilities on a similar scale to

Figure 5. The Russian connection: leaders of the two most advanced life support institutes in the Soviet Union visit Biosphere 2 in 1989. From left: Deborah Snyder, director of IE; John Allen, director of research and development, Biosphere 2; Yevgeny Shepelev, Institute of Biomedical Problems (IBMP) in Moscow; myself; Oleg Gazenko, IBMP; Linda Leigh, future biospherian; and Josef Gitelson, Institute of Biophysics (IBP) in Krasnoyarsk, Siberia.

the cyclotrons of the physicists. Considering what's at risk, the health of the biosphere and of humans, building facilities and laboratories to study global ecology in new ways is a crucial investment.

Ecotechnics: From Theory to Ultimate Test

Biosphere 2 would be a great testing ground and challenge for ecotechnics, the integration of the world of life with the world of technology. John Allen and about a dozen others, including myself, founded the Institute of Ecotechnics in New Mexico in the early 1970s. The word *ecotechnics* was inspired by technology historian Lewis Mumford. In *Technics and Civilization*, Mumford traces the evolution of human technics from its origins through the Industrial Revolution to modern times.[16] He called for the development of biotechnics, or technology supportive of human life, to replace impersonal "machine culture."

John Allen saw that technics needed to go further and be integrated with all life. He coined the word *ecotechnics*, where the goal is to integrate and harmonize *eco* and *techno*.[17] So we founded a small institute and started working on our first ecotechnic test case, the desertified Synergia Ranch near Santa Fe, New Mexico. We sought methods to combine ecological restoration and improvement with long-term viability, not short-term, profit-maximizing economics.[18] Now, four decades later, the Institute of Ecotechnics is registered in both the United Kingdom and the United States.

Part of my attraction to getting involved at our first location in New Mexico as a twenty-two-year old fresh out of college was that our work was not limited to ecology. In ecotechnic projects, we emphasize working in three areas: ecology, art (mainly theater), and enterprise. Enterprise allowed us to make ourselves and our endeavors self-supporting, a key component in these projects. I thought this offered a good model for what I wished for myself: personal growth and a balanced life. We took on bold and daring projects that expanded our capabilities through necessity. Our attitude was: *If it's already worked out, why do it?*

We investigated ecotechnics through hands-on field projects around the world, and built a research ship to sail the oceans. We also developed ecotechnic theory by convening workshops and conferences where cutting-edge scientists, managers, artists, and thinkers discuss significant issues.[19] IE, which I've directed since 1982, became a key consultant to Biosphere 2, drawing on our network for achieving the project's daunting challenges. Compared to our previous projects, where we relied on lots of sweat equity and a modest amount of seed capital, Biosphere 2 was orders of magnitude more complicated, expensive, and high profile.

Building Biosphere 2 would be quite radical; we had a whole world of unknowns to confront. One of our top consultants, Carl Hodges, director of the Environmental Research Laboratory at the University of Arizona, noted at a project review committee meeting where everyone listed their top ten concerns: "When you build a new world, you have all the problems in the world to solve!" So we knew it wouldn't be easy. But that's also why we thought the project could not fail. We would surely learn a lot and perhaps learn as much by what went wrong and what was unexpected as from what went right.

Beginning in December 1984, a series of meetings at the project site in Arizona brought together a distinguished and highly diverse group of scientists and engineers to consider whether such a project was even doable. We made the decision to go ahead with the Biosphere 2 project and created Space Biospheres Ventures to carry it through. In less than seven years, we completed research, design, and construction of the world's first biospheric laboratory. What was the next step? Eight biospherians had to close the airlock behind them and start the experiment.

2
Meet the Biospherians

I WROTE ABOUT my fellow biospherians in 1995 when the experiences were extremely vivid. After two years living in close proximity, I knew them as we rarely know other people. I could tell them apart by the tempo of their footsteps, their distinctive breathing. We were extraordinarily attuned to one another. Even on our two-way radios, with only a word or two I could discern their moods, ugly or sunny, focused or panicked, which was important—no, crucial—information, when working together was vital no matter what.

Eight Wildly Different Individuals

Here is the brave crew of biospherians. Each of us had a unique background. The team was chosen with their diverse array of talents and training in mind; hopefully we could take care of any problem that might arise in Biosphere 2.

Sally Silverstone was our solid and unflappable operations captain. She was thirty-six and was from a working-class area of London with a degree in social work from Sheffield College. Her education and training, along with an appreciation for the world's environmental challenges, were deepened by years of fieldwork as a social worker in Kenya and India. While studying tropical agriculture in Puerto Rico, Silverstone worked on the IE sustainable timber project in a secondary rainforest. Her Biosphere 2 responsibilities were crucial: She was manager of the agriculture and food systems. She measured out the rations for the cook and squirreled away harvests for later times of need. Gifted with a healthy appreciation for food, Sally supplied much of our limited quantity of specialty homegrown foods as a brewer of wine and beer, maker of cheese and sausage, baker of breads and mammoth-sized and creatively decorated birthday cakes. As co-captain in charge of day-to-day operations, she would call upon her mediation and negotiation skills to try to satisfy what eight highly individualistic people wanted to accomplish. Only in partial jest did Sally mention that her most important job qualification was managing a mental hospital, and that was nothing compared to eight biospherians.

Mark Van Thillo, thirty, lived up to his nickname "Laser," with his frenetic drive, energy, and straight-line intensity. Laser was co-captain with Sally and was also in charge of emergency situations. He had been responsible for quality control of the construction and installation of systems in the facility and was in overall charge of the technical systems inside. Born in Antwerp, Belgium, and of Flemish background, Laser spoke with a distinctly European accent. His lean six-foot frame gave him the look of a cowboy. A graduate of the Don Bosco Technical Institute, Laser was one of the first biospherian applicants and began his work with Biosphere 2 in 1984. His training, like most of the crew, involved periods aboard an ocean-going Chinese junk sailing ship, the RV *Heraclitus,* and work at ecological restoration projects in the remote outback of northwest Australia (IE's Quanbun Downs and Birdwood Downs) in the tropical savanna. On the *Heraclitus*, he rose to the position of chief engineer, keeping the vital engines and pumps operating. In Biosphere 2, the amount of machinery, piping, and equipment seemed endless; the water system alone had more than thirty different subsystems. Since Biosphere 2 was originally designed to be as self-sufficient as possible, Laser could use the workshop inside to fabricate replacement parts for virtually all the machinery should we run short of parts. He also had the foresight to include all sorts of spare pipe and steel just in case. During the two years, we would make use of those materials to improvise homemade solutions to the problems that arose.

Linda Leigh, thirty-nine, was in charge of the terrestrial wilderness biomes—the rainforest, savanna, thorn scrub, and desert areas. Born in Racine, Wisconsin, Linda studied botany at the University of Wisconsin and range management at Evergreen State College, where she earned her bachelor of science. Her fieldwork prior to Biosphere 2 was diverse: replanting native prairie grasses in the Midwest, studying elk populations and wolf introductions in Washington, and developing an ecological management plan for the Alaskan peninsula. She was in southern Arizona researching potential desert crops before joining the project, where she worked on design and acquisition of the plants and animals for the land ecosystems while training as a biospherian candidate. Taciturn and introspective, Linda often seemed most at home in the solitude of the wilderness. She had enjoyed the longest experiment in the small Biosphere 2 Test Module two years earlier, spending twenty-one days alone in one of the closed system trials prior to the completion of Biosphere 2. But that was easy for Linda, who loved being in nature and whose persona included a dash of eco-hermit.

Abigail ("Gaie") Alling, thirty-one, was director of research inside Biosphere 2 and was in charge of the marine systems—marsh and ocean. A graduate of Middlebury College, she has a master's degree from the Yale School of Forestry and Environmental Sciences. Born in New York City, her childhood was also spent at family homes in Maine and on the small islands off the Georgia coast. Her love for the sea developed early, and her work prior to Biosphere 2 included field studies of dolphins and whales in Greenland, British Columbia, the Antarctic Peninsula, Sri Lanka, and the Caribbean. Gaie combined effervescent personality and radiant smile with manic drive and steely determination when engaged. It had not fazed her when outsiders pointed out that no one had ever successfully created such a large living coral reef. Now, the coral reef was successfully installed, and waves lapped up on a sandy beach planted with coconut palms—in a glass house at an elevation of 3,900 feet in the Arizona desert! Adjoining the ocean was a replica of the Everglades marsh, with mangroves boxed up and transported across the continent. However, Gaie now confronted the possibility that all of that had been the easy part compared to the hazards the ocean and marsh faced when closed up in such an airtight box. The ocean coral reef was perhaps the most vulnerable of all our systems, but its guardian angel and protector was one of the strongest-willed biospherians.

Roy Walford, at sixty-seven, was the oldest of the crew. He was born in San Diego and educated at Cal Tech and the University of Chicago Medical School. The crew's physician in charge of biomedical research, Roy had been director of a research lab at the UCLA Medical School for some thirty years. His specialties were studies of the immune system and the aging process. A scientist in pursuit of the secret of eternal life, Walford pioneered studies of a low-calorie, high-nutrient diet. In mice, the diet slowed down the aging process, extended lifespan by 50 percent, and drastically reduced blood cholesterol. A strong believer in alternate careers, Roy had been involved for years in the worlds of experimental theater, avant-garde poetry, and performance art, his life sometimes combining adventure and science. When he was traveling around India, he convinced advanced yogis to let him insert a rectal thermometer while they were meditating so he could document their control of physiology.

Jane Poynter, twenty-nine, was born in Surrey, England. Her ecological background included working at the IE Australian cattle ranch and on board the *Heraclitus*, where she learned to dive and explore coral reefs. An early biospherian candidate, she had been on the project since 1985, working in the insectary (raising exotic insects), managing the prototype agriculture systems in the research greenhouses, and learning animal care. Inside, she was in charge of field crops and domestic animals. Jane has an upbeat personality—quick to laugh, with a whistle-while-you-work persona. Her English accent bespoke her upper-class background, in contrast to Sally's Jewish, working-class Cockney.

Taber MacCallum, the youngest of the biospherians at twenty-seven, was born in Albuquerque, New Mexico. His father was a radio astronomer at Sandia National Laboratories, his mother, from Australia, a psychiatrist, and rather than go on to college, he chose to begin work at Biosphere 2 in 1985. But his education was deepened by opportunities to help develop the unique analytic laboratory system that could operate without toxic chemicals and without nonrenewable chemicals. A space enthusiast, Taber had been selected by the project and graduated from the first program of the

International Space University, held at MIT, in 1989. He was very much the ambitious boy of the crew, managing the lab that included sophisticated gas chromatographs, mass spectrometers, and continuous sensors designed to check important and potentially toxic gases in the biosphere's atmosphere and water system. He also assisted Roy with biomedical research.

These were my seven extraordinary companions. At times I loved them— other times, not so much. I saw their high and low sides, and I'm sure they can say the same about me. I was the last of the eight selected, for I had been an alternate until six months before closure. It took me perhaps a millisecond to respond "Yes" to the opportunity to join the crew. At forty-four, I was the second oldest in the crew. Born in Brooklyn, New York, to first-generation Jewish immigrants from Russia and Poland, I had graduated summa cum laude in philosophy and pre-med sciences from Dartmouth College. Spurning the family tradition of becoming a medical doctor, I began developing new approaches to solving ecological problems at IE's first project site, Synergia Ranch in New Mexico. Over the next twenty years, I developed a pretty green thumb—for a New York City kid, that is. I planted more than a thousand trees in the high, semidesert at Synergia Ranch and helped regenerate overgrazed pastures in the tropical savanna of northwest Australia at Birdwood Downs. As one of the founders of the Biosphere 2 project, I had served as director of environmental and space applications, organizing our conferences at the project site as well as the international workshops on closed ecological systems and biospherics, and I coordinated the work of many of the outside scientists involved in design and research of Biosphere 2. Inside, my areas of expertise and responsibility included managing the wastewater recycling system (our sewage plant), the basement agriculture and fodder crops, and assisting Linda on operations and research in the land ecosystems. I was learning to be a scientific diplomat and usually had an outgoing personality. But I knew I had a tendency for angry choleric outbursts and at times could also be a world-class grouch.

For 731 days I saw my teammates at breakfast, lunch, and dinner. I partied with them. I labored in the fields and ecosystems with them. I ate 250 meals that each one of us cooked. We went in with great hopes of forming deep friendships. We came out with some of us barely on speaking terms. But as a team, we did what had to be done to keep Biosphere 2 operating. That reality was far bigger than our conflicts and personality clashes. Without a team of dedicated people willing to do whatever was required—putting in long hours, improvising, finding solutions for problems no one had anticipated—we would not have made it.

Selecting and Training the Crew

The eight of us had been selected to be the first crew from a larger group of fourteen biospherian candidates who were training and preparing at the Biosphere 2 site. The project had a potential cadre from whom to draw the team, people who had worked successfully at one of the IE-consulted projects and people working at Biosphere 2. The process involved self-selection. You had to *really* want to become a member of the crew to be considered as a potential candidate. There was also selection by the project's management. They had the final word on what combination of people would have the requisite set of skills and be more likely to succeed in operating Biosphere 2.

A memorable experience convinced me to throw my hat into the pool of candidates. In order to test engineering systems and gain experience with closed ecological systems, the Biosphere 2 Test Module had been built. Later, the key managers and biospherian candidates would spend twenty-four hours in the Biosphere 2 Test Module. My experience during that time, explored later in more detail, was profound. It was a leap from being immersed in the history and science of closed ecological systems to actually experiencing what it was like to be part of one. It was incredibly satisfying; my body, not just my mind, got it almost instantly. I knew from that moment I wanted to spend a prolonged time connected with a living system in such a remarkable way. I anticipated getting my chance in one of the later closure experiments in Biosphere 2.

Jane Poynter noted that "the training of the biospherians was unorthodox but very effective."[1] Part of the training was working either on the *Heraclitus*, at Birdwood Downs, or at Quanbun Downs stations in Australia. All were quite remote—the ship normally had a crew of ten to twelve people on voyages, and the Australian projects were two thousand miles from the nearest big city.[2] Working at those sites meant dealing with real-time situations with a small group of people. During these years, both the ship and the Australian projects were in tropical heat and required demanding physical work.

Jane commented, "Between life on the heaving sea and in the baked savanna, all biospherians underwent training in extremes. These were our rites of passage."

For Roy, the experience at three-hundred-thousand-acre Quanbun Downs in the outback country was clearly a shock. "You were just thrown in as another cowboy," he said. "Not a dude ranch, a regular working ranch. The cowboys would be Aboriginals—really rough drinking, hard-riding cowboys . . . You had to know how to ride, or learn to ride . . . and string barbed wire across the landscape and repair windmills and herd cattle and castrate cattle and get along with the Aboriginals and drive bulldozers. That was the psychological test."[3]

The biospherians also worked together during construction and operation of the research and development center. Training included work in a prototype agricultural system in the research greenhouses. After 1990, the crew operated the farm inside Biosphere 2 and learned to manage all the technical systems. Seven one-week trials were conducted before closure, with the crew living and working inside the facility. Biosphere 2 required people to manage key areas, with another crew member as a backup in case of illness. Roy and Taber received U.S. Navy emergency dental training, and Roy became licensed to practice general medicine in Arizona. In addition, several biospherian candidates did a one-hundred-hour introduction to medicine course at the University of Arizona College of Medicine and a three-day course in immediate treatment of trauma at the Georgetown University Medical School.

The idea that the biospherians just walked into Biosphere 2, like astronauts into a launch vehicle, was far from reality. As Gaie Alling noted in the film *Odyssey in Two Biospheres*:

> We had to think of common sense things. Okay, we're going to live in there for two years. What do we want to eat? Where is our water coming from? There was all this science and engineering going on, construction, but for us personally, to put ourselves in that situation, what water am I willing to drink,

what food do I want to eat for two years, and what am I going to be able to do in my leisure time, what's available to me? So it's not that someone else was building Biosphere 2 and we just walked in there. No, no. no. Not at all, we were all part of the biosphere in some aspect in great detail, in my case the marine systems and research in general. And we were also involved at a very personal contemplative level.[4]

The crew was organized like a ship or expedition. There were two co-captains: a day-to-day operational captain (Sally) and an emergency captain (Laser). The two other officer positions were research director (Gaie) and medical officer (Roy). Each of the eight of us were in charge of at least one major area: manager of terrestrial biomes (Linda), manager of food and agriculture (Sally), manager of farm animals and field agriculture (Jane), manager of marine biomes (Gaie), manager of the analytic laboratory (Taber), manager of the medical laboratory (Roy), manager of technical systems (Laser), and manager of communications (me). Farming was something we all did together, and in all areas there were other people involved as work demanded. We had at least two people trained in the use of the sophisticated equipment inside.

Credentials and Capabilities

Some criticized the project because the crew had few advanced scientific degrees, Roy being the only one with a professional degree. He was an MD and had been a research professor. Gaie had a master's in environmental sciences. She dropped going forward to a doctorate because of the opportunity to work with Biosphere 2 in her marine biology field. Linda had a background in botany and range management and years of ecological field research experience. Laser, Taber, Jane, Sally, and I did not have scientific training per se, but we had worked in ecotechnic projects where we helped manage real-life ecological applications in varied biomes.

I had also begun writing scientific papers and presentations, and I participated in scientific workshops and conferences representing the project.

As chairman of the Institute of Ecotechnics, I had organized our annual conferences. Some of the world's top scientists participated at these conferences, and they became part of the design, research, and advisory teams for Biosphere 2.

The idea that we should have assembled a team of eight PhDs, especially as the initial crew, seems almost incomprehensible. As Jane later wrote:

> Few of us inside Biosphere 2 claimed to be scientists—we were managers. In essence, we were capable technicians. But people generally see what they expect, and people expected that the crew of Biosphere 2 would be scientists or engineers. . . . The Scientific Advisory Committee (SAC) [which issued a report and recommendations about a year into our two-year closure] could have advised putting more professional scientists on the biospherian team, but did not, although they did recommend more 'practicing scientists with advanced degrees' on staff outside. This was principally, I feel, because they knew that Biosphere 2 needed more than scientists. The two-year experiment required people well versed in Biosphere 2's operations and its mechanical systems and well-trained technicians, which all of us were. Also they knew there were few people willing to incarcerate themselves for two whole years.[5]

We had a world to manage with a daunting amount of physical labor required, and we needed extraordinary adaptability to respond to whatever problems emerged. With only six years from the commencement of the project until the beginning of the first experimental closure, there was a need for a trained cadre of people. So project management drew from those who had demonstrated they overcame real-world problems in innovative projects.

When President Eisenhower created what would become NASA and announced that the United States was beginning space exploration, the program faced a similar challenge. Where to find people with plenty of experience and a can-do spirit? So NASA initially selected astronauts from the seasoned pilots in the military services. Later they enrolled civilians, including doctors, scientists, engineers, and technicians to undergo space training.

Our team also had an irreplaceable, intimate knowledge of certain systems that we had helped to design, install, and run. Our knowledge of Biosphere 2 was, in most cases, encyclopedic. We could visualize our world, from foundations to space frame superstructures, in great detail while remembering how every element got there. This was exemplified by Laser, who, as director of quality control for the construction from 1987 to 1991, had fly-specked the installation of every component of the vast technical infrastructure of the facility. Four of the crew wrote afterward:

> Because we had participated in its origin, [the biospherians] did not have to expend unnecessary energy at closure . . . to get the basic information about its internal systems or to learn how it was assembled. Because of this, we were there to learn about the system we had created and let the complex nature of this coherent closed system teach us about its dynamical cycles.[6]

Maximizing Research

We intended to make the initial closure as scientifically fruitful as possible. As part of the research and operational programs, there were more than one thousand sensor points distributed throughout the facility to automatically collect important data. The team was also scheduled to do detailed water and air analyses with the equipment in the analytic laboratory. Detailed histories of our health and physiology would be recorded by examinations in the medical lab.

Every plant in the wilderness biomes had been measured and mapped before closure with help from professors and graduate students at the Yale University School of Forestry and Environmental Studies. Similar mapping had been done in the ocean coral reef and marsh biomes. A bio-accessions program documented every individual plant and animal introduced from our research greenhouses and on-site quarantine greenhouses.

In addition, we had several research workshops at the site before closure. These kinds of workshops, like a genetics workshop organized by Dr. Steven O'Brien of the National Cancer Institute, expanded our initial research program, as did collaborations with researchers who helped design the biomes or who helped with specific issues. The project's Scientific Advisory Committee also networked us with interested researchers to help with problems that arose as well as instituting additional studies. There were at least a couple dozen research projects mapped out when we started. By the end of the two years, this list had grown to more than sixty scientific studies.[7]

We anticipated starting research programs to follow up on surprises we were sure Biosphere 2 would supply. Keith Runcorn, a member of our advisory committee from the University of Newcastle in the United Kingdom, played a key role in proving the then heretical science of plate tectonics and continental drift through his paleomagnetic studies. At an environmental symposium the day before closure, Runcorn pointed out why the community was excited about new scientific instruments and projects like Biosphere 2: You can't know what you will learn from the closure experiment because new scientific tools enable you to see what has never been seen.

After Biosphere 2, this approach would later be explained to me by my PhD committee chairman, H. T. Odum: Don't make your research proposal so detailed that you can't pursue what's most interesting during the course of your study. If you have all the answers already, why do the research?[8] In fact, as I told one interviewer, "Going into Biosphere 2, not only did we not have all the answers, we didn't even have all the questions."[9]

When I look back at the work we did in Biosphere 2, I don't think that if I had gotten my doctorate before going in that I would have been capable of doing more research and better coordinating with outside scientists to facilitate their studies. Jane also makes similar points:

> Some reporters hurled accusations that we were unscientific. Apparently because many of the SBV managers were not themselves degreed scientists, this called into question the entire validity of the project, even though some of the world's best scientists were working vigorously on the project's design and operation. The critique was not fair. Since leaving Biosphere 2, I have run a small business for ten years that sent experiments on the shuttle and to the space station and is designing life support systems for the replacement shuttle and future moon base. I do not have a degree, not even an MBA from Harvard, as John [Allen] had. I hire scientists and top engineers. Our company's credibility is not called into question because of my credentials: we are judged on the quality of our work.[10]

Guidelines for the second closure experiment, which started in 1994, followed the recommendations of our Scientific Advisory Committee. Specialist scientists would come in for shorter or longer periods to do research using airlocks to minimize air exchange. This way, we could accomplish research that a generalist crew member couldn't carry out while also taking advantage of the ten individual rooms in Biosphere 2.

During the second year of the first closure experiment, we also changed mission rules, as the SAC advised. We did bimonthly imports and exports using the airlock. New scientific equipment was brought into the facility, and samples were sent to outside labs. They had advised this change to increase the science accomplished and to lessen the crew's workload.

Preparing for Paradise, Bracing for Hell

We were well prepared and excited about being selected for the first crew. But this was tempered by concerns about another part of the experience: What would life be like for us, personally, and for us as a group of people? As I wrote:

How would we coalesce as a group? I had seen plenty of the tensions that can develop in small, isolated groups during my years in Australia. Interpersonal tensions can build to dangerous levels when humans are isolated and in intimate contact. There was also the problem of the "us-them" split—the rancor that can develop between the group inside and those outside. How would relations go with "Mission Control," the project's administrators, the engineers, scientists and outside staff on whom we would be dependent during the two years?

What kind of life might we invent for ourselves inside Biosphere 2? We were the first biospherians, Biosphere 2's first settlers and natives. Taber, Jane and Roy took in musical instruments; Sally her flute; Gaie and I had a book project lined up; Laser was going to master making documentary videos; nearly all of us had cameras; Jane had paints. Linda planned for her life as a botanist in a newly discovered world. Nearly all of us talked about a new

Figure 6. September 26, 1991. The biospherian crew on closure day entering the front airlock. From right: Gaie, Roy, Jane, Taber, Linda, me, Sally, and Laser.

relationship with nature that might develop from our close rapport and our dependence on Biosphere 2.

We each pictured our personal paradise, and braced for hell. Would we have the inner resources necessary? Would our world survive, flourish? We went in to a world that had been newly planted. The biomes were babies, the trees just emerging from the shock of being moved, the corals just recovering from the 2000-mile trip from the Yucatan. As far as we could tell, all systems were on go, and for the previous three months, the final countdowns had proceeded, inexorably leading to the doorway.[11]

3

A Technosphere that Protects and Serves Life

Noosphere and Anthropocene

THE DRAMA AT BIOSPHERE 2 started long before closure. From the earliest design workshops, it was clear the project needed great engineers and ecologists—but they rarely work together. Usually ecologists groan when they think about engineers. *Oh no, here come the bulldozers, cranes, and road- and bridge-builders; goodbye nature.* And engineers, well, they don't speak the same language as ecologists. Usually they don't have to consider from the earliest stages how their engineering and technical solutions will impact the surrounding ecosystem. It's no surprise then that our world is littered and degraded by the unintended consequences of our industry, technologies, and engineering works.

Vernadsky saw that humanity was now a very powerful force on planet Earth. Unlike any previous species, humans spread over the entire world. We use our ever-increasing ability to move matter to shape the world to our needs. Our towns, cities, and agriculture supplant wilderness areas, our mining and energy-extraction move extraordinary amounts of raw materials.

The task of our time, Vernadsky prophetically realized, was to integrate what he and his successors call the "technosphere" with the "biosphere." The technosphere includes all human activities (like technology and industry) and the two human-dominated biomes: agriculture and cities. He thought this integration was possible, foreseeing a "noosphere" (from the Greek word, *noos*, meaning mind or intelligence) developing as the next stage of Earth evolution.[1]

When Josef Gitelson, Oleg Gazenko, and Yevgeny Shepelev first visited Biosphere 2 in 1989, Gitelson said he had only one serious criticism of the project, but an important one: it had the wrong name and should be called Noosphere 1, not Biosphere 2. Great intelligence and sensitivity to the world of life would be required to successfully operate it. Today there is an effort to formally declare that our global biosphere has now entered a new geologic era, the Anthropocene (Anthropo, meaning humans), since human activity is the dominant planetary force.[2]

Getting ecologists and engineers to work together was essential to create a facility where the technosphere and biosphere were meshed in a healthy way.

Dissension in the Ranks

Discontent began when the engineers saw the preliminary architectural drawings by Phil Hawes and Margaret Augustine. What a waste of resources, they said. "We can save big bucks—and make construction and engineering so much easier—if Biosphere 2's design was simpler. Think big-box-store simple." Biosphere 2's directors answered with an unequivocal "No!" They were going to create the first man-made biosphere, and couldn't believe the engineers wanted it to be ugly and unimaginative.

Instead the architects created a stunning design that paid homage to world architecture: Barrel-vaulted space frames inspired by Babylonian forms, stepped pyramids like those of the Middle East and the Americas, and geodesic domes as a nod to the modern architectural masterpiece invented by Buckminster Fuller. Phil and Margaret were influenced by two of the greatest American architects: Frank Lloyd Wright and Bruce Goff. Given its vast following around the world and Biosphere 2's role in public education and inspiration, that decision seems absolutely spot-on. More than three million people have visited Biosphere 2.[3]

Figure 7. Margaret Augustine and Phil Hawes, co-designers of Biosphere 2, in front of a model of the facility. Augustine was also CEO of Space Biospheres Ventures (photo by Peter Menzel).

Project managers were mindful that Shah Jahan, who created the Taj Mahal and other architectural masterpieces of India, wound up imprisoned by his son for emptying the treasury. But the bean-counters in India have calculated that the Taj Mahal has enriched India many times over because people are drawn to its sublime beauty.

Living inside Biosphere 2, the crew delighted in the majestic architecture we inhabited. Had we lived inside a structure resembling a department store, it would hardly have been the same. The unimaginative structures that people work and live in have psychological impacts; so much modern architecture is inhumanly straight-lined and rectangular, made of the cheapest materials, repetitive and generic.

The Biosphere 2 architects made some important, subtle points. The human habitat tower housed the shared library and hosted parties and celebrations. It also gave the biospherians a place where they could look out at their world and the surrounding beauty of the project's location. Though the tower was sixty-five feet in height, you still had to look up to see the

Figure 8. The tower library at the top of the human habitat of Biosphere 2. The animal barnyard is lit up on the right side of the ground floor. In the rear are the barrel vaults of the agricultural biome.

highest point in Biosphere 2, the rainforest pyramid stretching five feet higher, reminding us that we humans are not the crown of creation—the ever-evolving totality of Mother Nature is vastly more powerful and occupies the throne.

Learning to Speak the Same Language

With architectural decisions made, engineering and ecological design proceeded over the next few years. Periodically ecologists and engineers got together at design workshops. Ecologists were confronted by their unknowns: You want to have lizards in the desert. What do they eat in their native habitat? What could they eat in Biosphere 2? Repeat those questions for every animal. It surprised me there was no ready answer to the exact impact each plant has on the atmosphere. Except for well-studied farm crops, there were no textbook solutions to how much CO_2 plants take from the atmosphere, how much they respire, or what chemicals their roots release in the soil. There's so much we still don't know about how our biosphere works.

The ecological team had to learn enough engineering to communicate their needs to the engineering team. For the stream in the rainforest *várzea* (lowland wetland area), fed by a waterfall and overflow from a pond, the engineers needed numbers: How much water will the stream hold, and at what rate should the water flow? Then they could determine the type and size of pumps required. The marsh modeled on the Everglades would have six zones, ranging from freshwater to salty coastal mangrove, with vegetation adapted to each. Then the engineers needed to know the concentrations of salt in each zone's water, and how fast the water should circulate between them. Then they could proceed to design systems to supply and maintain desired salinity and deal with contingencies like some waters getting too salty or not salty enough.

We joked at first that the ecologists and engineers couldn't insult each other if they tried. The languages they spoke, their methods, and their thought processes were too different. The size constraints of Biosphere 2 limited the ecological designers to creating "quintessential biomes." Their task, for example, was to figure out how to have the feeling of the vast horizons characteristic of a savanna landscape in a small area, how to replicate the sensory feel of being in a moist, shady rainforest or fragrant, arid desert. Then the engineers—whose language was quantity and process—were tasked with figuring out how these challenging designs could actually be built, operated, and maintained.

Life and Technology Merge

Biosphere 2 was a world where life and technology met—and in a profound way, merged. This was a radical contrast to our planetary biosphere where our technologies are waging war against our biosphere. Inside Biosphere 2, technology was needed to replace some functions that the global biosphere performs. Without these technological assists, none of the life inside Biosphere 2 would have survived. Computer controls and irrigation devices replaced rain. Heating and cooling systems were substituted for the global

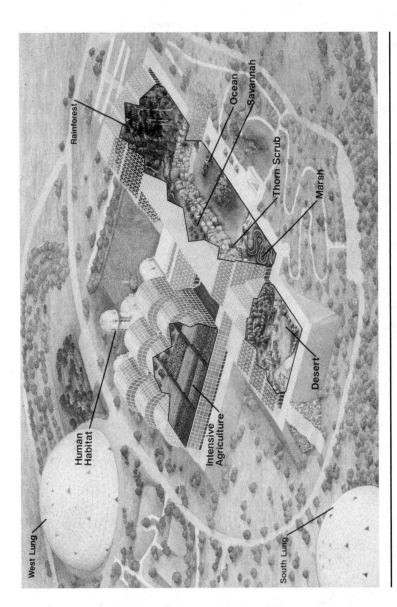

Figure 9. Schematic of Biosphere 2 showing the layout of the wilderness biomes, farm (intensive agriculture), human habitat, and lung structures. All of Biosphere 2 shared water resources and atmosphere. The lungs are connected to the main structure by underground air tunnels (drawing by Elizabeth Dawson).

climate system. Desalinators and pumps assisted evaporation and gravity in getting water in the right quality and quantity to where it was needed.

Kevin Kelly, founding executive editor at *Wired* magazine, sees this fusion of life and technology accelerating in the future. Technology will become more intelligent and adaptable, and we humans will get better at ensuring our technosphere will enhance and won't act at cross-purposes with the world of life.[4]

Biosphere 2 prioritized life over technology. An early confrontation dramatically illustrated this. Engineers alerted to a malfunctioning pump went in and hastily hacked back vines that had grown into the device. The ecologists were outraged at the damage to one of their carefully chosen plants. Mediation resolved matters and helped the engineers realize life took precedence in this new world.

Engineering Challenges

The engineers faced formidable challenge after formidable challenge.

Task number one: seal the structure so there would be virtually no air exchange with the outside. Bill Dempster, who coordinated systems engineering for Space Biospheres Ventures, took the lead. He worked with Peter Pearce, who had collaborated with Buckminster Fuller on the science of synergetics and whose company made our space frames. They stepped in after our consulting engineers tried several conventional methods, which proved far too porous. The roof of the test module was the test case, and finally, after total failure, it was removed by crane. Engineering began again. Finally, Dempster and Pearce came up with a new patented design to build the sealing into the engineering of the space frames. Underground, Biosphere 2 had walls and floors of welded, high-strength stainless steel. Just concrete would not have made the structure airtight enough. Overhead there were more than twenty miles of space frame joints to be sealed.

Dempster designed a sophisticated system to find tiny leaks so they could be sealed. An underground tunnel was built around Biosphere 2 where inert trace gases released inside the facility could be detected. This not only determined the leak rate but showed which sector of the stainless steel liner was

Figure 10. Team of glaziers working on the glass space frames. They installed more than six thousand frames, each six feet per side. This work was often done at night because of blazing daytime temperatures. Many of the space frame crew were rock-climbing mountaineers (photo by Peter Menzel).

leaking. Many were found and sealed. The leak rate finally achieved for the two-year experiment was less than 10 percent per year, which amounted to less than 1 percent per month. That's the equivalent of all the tiny holes in liner and space frame adding up to one hole the size of a pinkie finger.[5]

But if the structure was that airtight and covered with glass space frames to allow in as much sunlight as possible, what happened when the air pressure inside differed from the outside air pressure? Answer: the structure would explode if inside air pressure was higher, and it would implode if it was lower. This is because air expands in volume when it's warmer and contracts when temperatures fall.

Bill Dempster again came to the rescue with an out-of-the-box solution. He suggested allowing part of the facility to expand and contract. He and the engineers designed two variable volume structures attached through an air tunnel to Biosphere 2. When inside temperatures rose, the force of the expanding air pushed up a rubber membrane weighing fifteen tons. At

Figure 11. Inside one of the Biosphere 2 lungs. The pressure of the air is lifting the rubber membrane, which weighs some fifteen tons, off the floor. The lung membrane moves up and down as air pressure and volume change.

night, when inside temperatures got colder and air contracted in volume, the rubber membranes fell. The difference in height could be as much as twenty feet.[6] Because the two membranes went up and down, the two variable volume structures came to be called "lungs." But unlike our human lungs, they did not take in and release air. They allowed the internal air to expand and contract without damaging the structure. The Biosphere 2 lungs allowed us to have a lightweight structure using space frames with glass to let in sunlight.

The engineers had to carefully consider the outgassing potential of all material or equipment and the operating and maintenance time they required. Several biospherian candidates worked on the Biosphere 2 construction. Laser, in charge of quality control, helped ensure that systems wouldn't be too difficult or time-consuming to operate and maintain.

Vacuum Pumps for Waves

There were other engineering concerns. Coral reefs and marine ecosystems require water movement for nutrient circulation and filter-feeding. Meetings between ecologists and engineers got stormy about pumps to create the ocean waves. The marine ecologists were terrified about the death of their creatures. Normal mechanical pumps would take in phytoplankton and fish and kill them. The engineers came up with a solution—vacuum pumps. These pumps sucked up ten thousand gallons from the million-gallon ocean and gently let the water fall. The vacuum pumps required a wall between ocean and marsh, with a cavity where this volume of water could be vacuumed up and then gently released.

Linda Leigh compared the noise the vacuum pumps made to the sounding of a large gray whale. I heard it as the recurring in-breaths of the "old man of the sea." It became an integral part of the soundscape we biospherians listened to for two years. Every twenty to thirty seconds, a wave was released. You could hear it throughout Biosphere 2. Not hearing that sound meant there were pump problems—instinctively alarming to all of the crew, like sailors accustomed to the sound of ship engines. When the wave machine stopped, this triggered alarms and emergency response since our ocean's health depended on it.

The force of the waves dictated the shape of the beach. An early disaster was the erosion of beach sand by a too-powerful wave. A breakwater was installed to protect the sand, and waves were made gentler. Even the choice of sand was dictated by the needs of life. Sand from the Gulf Coast was trucked in, as local sand was too sharp for the feet of amphibian species.

The glass used in the space frames excluded almost all ultraviolet radiation. The cost of using special glass was prohibitive, so the ecologists had to exclude daytime active lizards in the biomes since they require UV rays for making vital nutrients. We also took supplemental vitamin D tablets, because without UV light, our bodies couldn't produce their own supply.

Engineering a Climate

A closed ecological system means as little material exchange between inside and outside as possible. But all ecological systems must be open to energy flow (for electricity, cooling and heating, and, in our case, sunlight) and the exchange of information.

Seasonal sunlight length was important since sunshine powered Biosphere 2's plants. In some low-light areas of the facility, supplemental electrical lights were installed. The engineers had to cool and warm the facility. There was no worldwide climate produced for free by nature, as there is in our global biosphere. Cooling was the number one concern and required the most energy, as expected for a greenhouse in southern Arizona. Without cooling, within a few summer hours, it would reach 150 degrees Fahrenheit inside Biosphere 2, killing every green plant inside. Therefore, the engineers installed backup systems—safety through redundancy. In the energy center outside, a natural gas generator was the main supply, chosen because it was the least polluting fossil fuel, with a diesel backup.[7] Solar energy was expensive in the 1990s. Using solar would have increased project costs by $20 million. The generator's waste heat was used to reduce overall energy demands by at least 10 percent. Excess energy generated fed into the power grid for the region. Biosphere 2's energy center kept the lights on in a neighboring town for several days when their grid power supply was being upgraded.

We needed to heat and cool Biosphere 2 so the inside biomes remained within required temperature ranges. For example, the rainforest shouldn't get above 95 degrees Fahrenheit in the summer nor fall below 55 degrees in the winter. In the desert, temperature ranges could be more extreme, with summer heat allowed to reach 110 degrees and dropping into the 30s in the winter. We chose a fog coastal desert adapted to higher humidity because

Table 1. The energy system of Biosphere 2. Cooling was the largest demand on the energy system. Solar energy powered the growth of the green plants inside.

Energy	Peak	Average
Electrical	500 KW	200 KW
External Support	2,000 KW	1,200 KW
Heating	11×10^6 kj/hr	5×10^6 kj/hr
Cooling	30×10^6 kj/hr	10×10^6 kj/hr
Solar flux	27×10^6 kj/hr	6×10^6 kj/hr
Photosynthetically Active Radiation (PAR)	30 moles m^{-2} day^{-1}	20 moles m^{-2} day^{-1}

it was too energy-demanding to dehumidify. Biosphere 2's wilderness wing had high humidity since it included a rainforest and ocean. But in the human habitat and farm, for comfort and to reduce pest infestations, the humidity was kept far lower. The farm also had tighter temperature regimes to maximize the productivity of our mix of tropical and temperate crops.

These varying temperature ranges required sophisticated engineering. Closed-loop water pipes warmed Biosphere 2 when needed during the winter and removed excess heat during the rest of the year. The heated and cooled water flowing in those pipes was not actually part of Biosphere 2's matter, but the pipes were. The water was a method of creating energy flow, for inserting and removing heat and cold.

Life Calls the Shots

I grew to admire our staff and consulting engineers. They were smart, always resourceful in evaluating issues and inventing technical solutions. Engineers pride themselves on the safety and reliability of their systems. But another wave of discontent arose from some of the engineers. They were learning how to work with the ecologists, and vice versa. But they realized, despite brilliant innovative engineering, they couldn't guarantee that Biosphere 2 was going to be livable and healthy. That depended on the life of Biosphere 2, especially the microbes, fungi, and plants.

The same is true, of course, in our global biosphere. It's the life on Earth that cleans our water and air and produces our food. Without green plants there wouldn't be 21 percent oxygen in Earth's atmosphere. Without life, nutrients wouldn't be recycled to sustain and power more life; crucial biogeochemical cycles would grind to a halt. But unlike our planetary biosphere, which thrived for billions of years before humans evolved and will likely still carry on if we disappear, Biosphere 2 was fundamentally different. We humans were vital to keep all the engineering systems and machinery operating that supplied functions in Biosphere 2 that nature does outside so reliably without any outside help. But without the action of life, Biosphere 2 wouldn't have breathable air, drinkable water, or safe food despite all the elegant and intricate engineering.

This was shocking to the engineers. They ensure the safety and well-being of people. Engineers make sure roads are safe, buildings and bridges don't fall down, airplanes fly. But now, in Biosphere 2, the balance of power had been changed. Life was driving everything, making the decisions, dictating what materials, equipment, and systems would be installed. Life was number one, the technosphere was there strictly to serve and protect *life*. *Life* was the overriding factor—not profit or expediency or time. What promoted and facilitated the flourishing of life—from microbes to large vertebrates like humans—was the deciding factor for every engineering decision, choice of material, equipment, or system.

After the shock passed and the engineers fully understood their crucial supporting role in helping the biosphere work, we had some of the highest-morale engineers anyone has had the pleasure of working with. They loved solving intractable problems and pushing the envelope. They loved engineering with the total context deeply in mind. They got it—everything they proposed *had* to be supportive of and not injurious to the life of Biosphere 2.

This was quite a revolutionary idea, though you might think it's simply common sense. Why should we do the same outside? Our global biosphere is far, far more tightly sealed than Biosphere 2. The Earth loses a tiny amount of light gases like hydrogen and helium from the upper atmosphere, and some cosmic material enters in the form of asteroids and meteors. The net loss is far less than one-trillionth of a percent per year.[8] Everything else circulates and cycles. There is nowhere to dispose of toxic chemicals or pollution permanently.

Our Addiction to Toxic Chemicals

Barrels of chemicals and stored toxic waste washing from warehouses were among the most frightening images I saw in Biosphere 2 from television coverage of the Mississippi River floods in 1993. From my perspective in a world we worked so hard to keep healthy, I shuddered contemplating the madness of industry's addiction to synthetic chemicals. The scariest thing was not that many of them are known to be toxic and carcinogenic, but that there was—and is—rarely any safety data supplied or testing carried out for new chemicals before they are used in industry.

The United States EPA has safety tested only a small percentage of the eighty-five thousand industrial chemicals in use. Only five chemicals have ever been banned or restricted.[9] Five trillion pounds of industrial chemicals are produced worldwide each year. Every human now has hundreds of synthetic chemicals in their body; because they are in their mothers' bodies, even unborn babies have them. Exploding disease rates (including cancer, asthma, developmental disorders, ADHD, and autism) may be linked to chemical pollution of our world.[10]

The warehouses images underlined the fantasy of thinking we could safely store dangerous chemicals so they would not get into the biosphere's cycling machine. Testing chemicals before they're released would be an excellent use of minibiospheric laboratories or far smaller ecological mesocosm (modular biosphere) facilities.[11]

The precautionary principle for chemicals would mean erring on the side of safety when human or environmental health was threatened rather than finding out later they caused harm.[12] If chemical companies and industries bore the burden of proving their activities were harmless, this would catalyze the search for safer alternatives.

We humans are just starting to get our minds around the concept of the biosphere and the implications of its closure.[13] In Biosphere 2, the rapidity of the cycling and smallness of its volume made this an inescapable reality. There was no room for denial. It's just that Earth's size and weight are vastly greater, its cycles longer than those in Biosphere 2. An ancient proverb says, "Don't foul your own nest." It's high time we realize that our nest is Earth's entire biosphere.

Engineers would rise to the challenge if this was the requirement for everything they are asked to accomplish in the planetary biosphere. We must challenge engineers, industrialists, and chemists to redesign industry to meet our needs without poisons and dangerous chemicals, to develop methods to clean up existing chemical pollution. We must make our global technosphere serve and not harm life, ours and all life in our biosphere. We just don't ask nor require them to do so. Yet.

4

"Every Breath You Take, Every Step You Make"

A Small, Life-Packed World

VISITORS TO BIOSPHERE 2 are impressed by how large a structure it is. We biospherians were acutely aware of how small and finite it was. Yes, space frame ceilings were some seventy-five feet above the ground in the rainforest, mostly forty feet elsewhere. The wilderness wing—rainforest, savanna, thorn scrub, fog desert, mangrove marsh, tropical ocean with coral reef—stretches more than five hundred feet, almost the length of two football fields. That, plus the farm and living areas and two nearby lungs, makes the overall footprint a bit more than three acres. Yet each one of our biomes was scarcely half an acre. You could get across any of them in a few minutes at a slow walk. Compare a seventy-five-foot ceiling with Earth's atmosphere, extending fifty *miles* above the planet.

Our rich soils to grow the young trees we started with to full height extend four feet deep in the farm and eight to ten feet deep in the wilderness biomes. Over a couple of acres that adds up to thirty thousand tons of living soil.[1] The Biosphere 2 ocean was also packed with life, supporting the world's largest man-made coral reef and dozens of tropical fish species. Our ocean didn't have the life-scarce stretches of deep, nutrient-poor ocean waters, called "ocean deserts." Sunlight fully penetrated the Biosphere 2 ocean since it was only twenty-five feet deep at its deepest end, powering the growth of algae and other phytoplankton and the plant part of the coral symbiosis.

Each living world has its unique rhythms. In our global biosphere, CO_2 molecules remain in the atmosphere for three years, on average, before they are taken in by a green plant. That's its residence time. But make a new world—a very small, closed ecological system, or Biosphere 2—and the rhythms and cycles are quite different. In Biosphere 2, we calculated that CO_2 residence time in our tiny atmosphere was two to four days. In a year, CO_2 completes 90 to 180 cycles, about 250 to 500 times faster than in our relatively spacious global biosphere.[2] Water cycles and nutrient cycles accelerated as well. John Allen, inventor of Biosphere 2 and its first director of research and development, called it a "cyclotron of the life sciences."[3] That's great for research, but a nail-biter for management. Things happen fast in a minibiosphere.

Living with these accelerated cycles, we biospherians joked we'd entered a time machine. Our exploration ventured into altered time scales rather

Figure 12. Lightning over the Santa Catalina Mountains, with the "cyclotron of the life sciences" in the foreground (photo by Peter Menzel).

than traveling to distant lands. Gaie, in particular, thought our adjustment during the two years was attuning ourselves, getting our bodies in resonance, vibrating in accord with a faster-tempo world. She estimated it took about eighteen months to be perfectly at ease in Biosphere 2 and about twenty-four months to fully readjust to living outside.[4]

Sick Building Syndrome

We were terrified of the buildup of harmful trace gases, which would result in "sick building syndrome." The superb engineering of Biosphere 2 achieved an air leak rate of less than 10 percent per year, far tighter than the International Space Station and thousands of times more sealed than a modern home or office building.

But, there is an invisible world of molecules. We have the illusion that solids are, well, solid. Yet everything releases gases. That process is called outgassing. Animals, plants, synthetic materials, machines, humans, paints, glues, solvents, and computers all outgas. There are biogenic gases (from living organisms), technogenic gases (from man-made materials and equipment), and anthropogenic gases (from humans).

In NASA's Skylab space station and shuttle, hundreds of trace gases accumulated in the cabin atmosphere despite all the precautions to carefully screen materials and equipment to minimize outgassing.[5] For space applications, charcoal or other filters and catalytic oxidizers are options to reduce trace gas buildup. All these fixes require substantial energy or expensive consumable parts.[6] These approaches wouldn't work in the large atmosphere of Biosphere 2. We needed to find a different solution.

Trace gas accumulation may become unnoticeable to those inside. A space scientist shocked me with his story of being among the first to greet the returning astronauts when they landed the space shuttle. He took a whiff and promptly threw up. Though the shuttle was only in space for a couple of weeks, the interior air was absolutely disgusting—dank, stale, revolting. Often people in such closed spaces don't realize it. A few years later, at a space conference, Yevgeny Shepelev confided to me that after the jubilant cheers of his fellow researchers, he went back to the chamber where he'd

Figure 13. Taber MacCallum, collecting air samples in a wilderness biome (photo by Abigail Alling).

spent twenty-four hours with chlorella algae. He said it stank inside, and he was amazed he had endured the experiment.

Energy-efficient and tightly sealed modern buildings and homes are subject to trace gas buildup, which can cause illness. Since Biosphere 2, this problem has gained considerable attention. Some obvious culprits, like new carpets made of synthetic materials, must now vent their trace gases for a sufficient time before installation.

Everything that went into Biosphere 2 was scrupulously screened to avoid those with dangerous outgassing. Wool and wood was used extensively for flooring, wall paneling, and furniture inside the human habitat, which housed our kitchen, laboratories, personal quarters, and workshops. But Biosphere 2 was a giant apparatus, with hundreds of motors and pumps. Other equipment included desalinators and twenty-five air-handling machines that circulated air creating mild breezes. Numerous tanks stored different quality water. Miles of closed-loop pipes brought in heated water in the winter and removed heat in the summer. Personal and workstation computers were distributed throughout the structure. Roy joked that "Biosphere 2 is like the Garden of Eden on top of an aircraft carrier."[7]

Air and water quality was such a priority that the project developed an innovative analytic laboratory. Unlike ordinary labs, it required no toxic chemicals or nonrenewable consumables. This enabled precise, real-time monitoring so we could take corrective action. The lab also enhanced our research program. The Gas Chromatograph Mass Spectrometer (GCMS), for air, and the Ion Chromatograph, for water, detected trace gases or water impurities in parts per billion, sometimes parts per trillion.[8]

In addition to manual sampling and testing, two automatic systems detected air and water pollution. The "sniffer" monitored the air, and the "sipper" monitored the water. The sniffer provided continuous readings on eight trace gases. The sipper monitored critical water parameters in the ocean like pH levels (acidity), dissolved oxygen, and temperature.

Microbes: The Unsung Heroes of the Biosphere

Ingenious use of our soils and plants provided our trace gas solution. The method uses a soil bed reactor, or soil biofiltration. German engineers in the early twentieth century had the then-unheard-of idea to fix nasty smells coming out of factories and sewage treatment plants by routing the exhaust pipes through the bare soil of the surrounding land. The air loaded with the odor-producing trace gases was exposed to the staggering diversity and numbers of soil microbes.[9]

In ecology, not only is everything connected to everything else, but everything is part of a food chain. Everything eats and is eaten (or decomposed) by something else. So, it's no surprise that the trace gases we find objectionable or potentially dangerous are food for some group of microbes. Almost without exception, there are microbes that metabolize, consume, and make harmless whatever trace gas is causing a problem.

Luckily, our team met Dr. Hinrich Bohn, a University of Arizona professor from Germany, where soil biofiltration is far more widely used than it is in the United States. We experimented to see if growing crops could combine

Figure 14. The Biosphere 2 Test Module, a research chamber used from 1986–1991 to test sealing, engineering components, and run closed ecological system experiments, including three- to twenty-one-day human closures. View is from its variable volume chamber. It is still one of the largest closed ecological systems in the world (photo by Gill C. Kenny).

with soil biofiltration. Carl Hodges and the Environmental Research Laboratory (ERL) he headed served as prime consultants to Biosphere 2 on agriculture and other engineering aspects of the project. Seventy-two plant containers were tested at ERL. A pump blowing air up through the soil slightly improved crop growth, probably because of better aeration of the root zone. This marked the first time soil biofiltration included growing plants.

A series of experiments at ERL and in the Biosphere 2 Test Module confirmed these plant and soil biofilters reduced trace gases. We'd spike the atmosphere with a measured quantity of a trace gas such as methane, carbon monoxide, sulfur dioxide, or ethylene. It took part of a day before the soil microbes able to consume the gas increased their numbers. Then, reveling in the feast we'd provided, those microbes rapidly lowered the trace gas concentrations to safe levels.[10]

From the first Biosphere 2 design workshop, we'd known that microbes would be vital to making our world healthy. We called microbes the "unsung heroes of the biosphere." They completed the cycling of vital life elements and kept water and food healthy. Microbes ran our global biosphere for more than two billion years before multicellular organisms evolved. Recently scientists have learned much more about the microbes inside us. The human microbiome comprises microbes, fungi, and viruses. They enable us to digest food, supply energy, make vitamins, and help fend off disease. We are not alone—there are about as many microbial cells in your body as human ones.[11] There are five hundred to one thousand species of bacteria living in our guts and on our skin.[12] We are all ecosystems.

Soil biofiltration was incorporated in Biosphere 2. The entire farm soil (our agriculture biome) was designed to be our soil biofilter. Air pumps below in the technical basement could, if needed, push the entire atmosphere through the soil in twenty-four hours. We didn't use the soil biofilter during the two years because all the green plants and passive infiltration of the air into the soils succeeded in keeping all our trace gases in check. Nitrous oxide was the only exception, since this trace gas is only decomposed in the stratosphere by high-energy ultraviolet rays.

A test of the soil biofilter in Biosphere 2 caused an increase in CO_2 because soil air has far higher concentrations of the gas than the atmosphere. Our concern with CO_2 management gave another reason not to use it during the experiment. In situations when it's desirable to boost CO_2, activating a soil biofilter could be a good strategy.

Dr. Bill Wolverton, then of NASA's John C. Stennis Space Center, pioneered research showing that plants help clean up indoor air. Air purification can be fifty times more effective when the plant also has its soil activated by an air pump. Wolverton also helped us design constructed wetlands for wastewater recycling.[13]

Space frame roofs in Biosphere 2 were high both for tree growth and to make our atmosphere as large as possible. But despite seven million cubic feet of volume, Biosphere 2 was extremely susceptible to trace gas accumulation.

Our GCMS showed soon after closure all but two of the detected trace gases were rapidly declining in concentration. Construction work was intense right before we entered, with engineers and technicians completing myriad systems. Therefore, many technogenic gases had increased in the air. These two trace gases were still slowly rising instead of declining like all the others. Consultants advised these gases were from PVC glues and solvents. Eight of us fanned out searching for the culprit. Finally, in a dark corner of the technosphere basement, we found a couple of small, improperly closed cans of glue and solvent. Their caps were cross-threaded. Those tiny air passages allowed trace gases to escape and accumulate in measurable quantities. The cans were properly tightened and, within a week, concentrations of those trace gases declined substantially in our atmosphere.[14]

Atmospheric Vigilance

From the moment the Biosphere 2 airlock doors closed behind us, breathing was never taken for granted. With such a small atmosphere and such a large technosphere, we had to be vigilant about any action that might pollute the air we breathed. Each morning meeting covered atmosphere reports, including CO_2 and oxygen levels, as well as trace gas levels from periodic GCMS tests. Irrigation changes or improvements often required cutting and gluing PVC pipes. We evaluated whether it was safe to release the trace gases associated with cleaners, solvents, and glues. These discussions could get heated, but sometimes we had to wait until trace gas levels were lower.

I wonder what would happen if residents of a city saw a complete analysis of the air they breathe. Air inside homes and offices can be fifty to one hundred times worse than outside air.[15] Daily or weekly publication of exactly what you're breathing would raise awareness of this crucial health issue. We already have a smog index in many cities to alert people to grosser air quality issues. What if city residents had a voice in deciding what kind of technologies and developments should be allowed, based on their impact on their air quality?

More than seven million people die prematurely every year from air pollution. Bad air is strongly linked to strokes, heart problems, and cancer, as well as acute respiratory diseases.[16] Polluted air is now ranked as the largest environmental health hazard, causing one-eighth of total world mortality.

Figure 15. Biospherian morning meeting over breakfast. Far side of table from left: Jane, Linda, Roy. Front left: Taber, Sally, author. Left head of table: Laser. Right head of table: Gaie (photo by Abigail Alling).

This number may double in coming decades if industry and agribusiness farming expand without better technologies and regulation.[17] Acid rain, depletion of the ozone layer, and rising greenhouse gases causing climate change demonstrate we share our atmosphere with everyone and every living creature. What happens elsewhere in the biosphere impacts us. It may be self-calming, but, in fact, it's delusional to think that air pollution just *goes away*. The reality is captured in the saying "What goes around, comes around," in small or large closed ecological systems like Earth.

We should let nature clean up our air. Remember the taste of the air when you're in a forest or beautiful green park. If your home or office doesn't have green plants growing in a rich soil, it's not only less pleasant, but our natural allies do not have a chance to do what they do so well elsewhere in the biosphere—keep the air healthy. Greening cities, homes, and offices improves many things, including the air we breathe.

Soil biofiltration was one of the first environmental technologies we developed at Biosphere 2 for commercial application. Our engineers designed indoor plant containers, called "airtrons," that functioned as a soil biofilter. Below the plant pot is an air pump that forces the office or home's air up through the soil, greatly increasing its ability to cleanse the air of trace gas

Figure 16. Putting nature to work, an "airtron," or soil biofilter. Below the plant is an air pump, so the indoor air interacts more thoroughly with the soil. In the photo is Anita Cota in the Biosphere 2 Mission Control building.

contaminants. The beauty of putting nature to work is that the soil microbes will respond to whichever trace gases are present. They don't need to be told what to do![18]

As we stepped inside Biosphere 2, we left the thin, dry air of the high desert. One breath, and we sensed the dense air of our life-packed new world—rich with fragrant tropical aromas. We thought carefully before we did anything that might pollute our air and jeopardize our safety.

5

Carbon Dioxide

Atmospheric Management of a Small World

Living Inside a Cyclotron

THE RISE OF CARBON DIOXIDE in the Earth's atmosphere was not on everyone's minds at the time of our biospheric experiment. Inside Biosphere 2, it sure was. Managing CO_2 was a major concern and daily hot topic. It could literally be a showstopper, but not because of global warming. In Biosphere 2, our temperatures and rain were controlled by our technosphere, but extremely high levels of CO_2—or worse, runaway CO_2—would be disastrous.

The ratio of carbon in living biomass (all the organisms) to the amount of carbon in our atmosphere was a hundred times greater than on Earth. The ratio of our soil carbon to atmospheric carbon was more than two thousand times greater.

Scientists and consultants had warned that this was going to be a tough problem to solve, raising the disturbing question: Could we solve it?

Global warming has reminded us the atmosphere is far from static. Over vast stretches of geologic time, the atmosphere has dramatically changed. Our planet's early atmosphere was nothing like what it is now: There was no oxygen in the atmosphere until life invented photosynthesis more than three billion years ago. Carbon dioxide has greatly fluctuated, dropping during

Table 2. Estimated carbon distribution on Earth and in Biosphere 2. The ratio of soil carbon to atmospheric carbon is 2:1 in Earth's biosphere and 5,000:1 in Biosphere 2.

	Earth	Biosphere 2
Soil mineral component	94.88%	8.9%
Marine mineral component	5.1%	5.6%
Marine organic sediment	—	1.7%
Soil organic matter	0.006%	82.3%
Plants	0.003%	1.5%
Atmosphere	0.003%	0.015% (at 1,500 ppm)

Sources: B. Bolin et al., "Interactions of Biogeochemical Cycles," in *The Major Biogeochemical Cycles and Their Interactions* (New York: Wiley & Co., 1983); M. Nelson et al., "Atmospheric Dynamics and Bioregenerative Technologies in a Soil-Based Ecological Life Support System: Initial Results from Biosphere 2," *Advances in Space Research* 14, no. 11 (1994): 417–426; M. Nelson et al., "Using a Closed Ecological System to Study Earth's Biosphere: Initial Results from Biosphere 2," *BioScience* 43, no. 4 (1993): 225–236.

ice ages and rising when the Earth became warmer and more tropical. We made a miniature biosphere with sharply increased ratios of carbon and far faster cycles. We didn't expect it would settle down and immediately form a perfect balance.

Our carbon cycling was hundreds of times faster than in our global biosphere. Large amounts of carbon in soil and biomass, coupled with a very small atmosphere, meant Biosphere 2's carbon atmospheric residence time was just a few days compared to three years in our global biosphere.[1] We had 90 to 180 cycles of carbon every year, rather than one-third of one cycle. Two years in Biosphere 2 would equal hundreds of years of Earth's atmospheric carbon cycling. It became another dimension of our time machine; we weren't traveling anywhere, but cycles were accelerated like in a sci-fi movie. Our world hummed with faster metabolic rhythms. A folk saying distinguishes "the quick and the dead." The quick refers to the living. In Biosphere 2, it was quick, quick, quicker; the life pulse was greatly accelerated.

We ran test module experiments with a person inside for three, five, and twenty-one days. Carbon dioxide, while far higher than in the outside

Table 3. Ratios of carbon in key reservoirs and the resulting speedup of the carbon cycling time in Biosphere 2.

	Earth	Biosphere 2
Ratio of biomass carbon: atmospheric carbon	1:1 (at 350 ppm CO_2)	100:1 (at 1,500 ppm CO_2)
Ratio of soil carbon: atmospheric carbon	2:1	5,000:1
Estimated carbon cycling time (residence time in atmosphere)	3 years	2–4 days

Source: M. Nelson et al., "Earth Applications of Closed Ecological Systems: Relevance to the Development of Sustainability in Our Global Biosphere," *Advances in Space Research* 31, no. 7 (2003): 1,649–1,656.

atmosphere, did not rise enough to become a health concern. CO_2 also rose during weeklong experiments inside Biosphere 2 prior to the full closure experiment. But because there were so many other people working inside, it was hard to extrapolate from that data. Our life systems were young. We expected rapid growth not only over the first two years, but over the early years of Biosphere 2's planned hundred-year lifetime. As the plants grew, there would be more resources to manage the atmosphere.

But we started with a baby biosphere. Could it cope? Some very smart scientists, inside and outside the project, predicted there would be a runaway rise in CO_2 and an early end to our two-year experiment. John Allen called carbon dioxide the "tiger" of Biosphere 2—a force of nature, potentially extremely damaging if it spiraled out of control. It was our first big challenge—and if we lost, our last.

CO_2 concentration was 280 ppm (parts per million) in Earth's atmosphere before the Industrial Revolution began in the 1800s. Ever since, it has continued to rise, recently exceeding 400 ppm. Other greenhouse gases such as methane and sulfur/nitrogen oxides have also risen, triggering deep concerns about the impacts of global climate change.

For human health, this rise in CO_2 is not a concern since we can tolerate far higher concentrations. The space shuttle typically fluctuated between 5,000 and 10,000 ppm (which is 1 percent CO_2). The International Space Station operates between 2,000 and 7,000 ppm without health concerns,[2] and submarines operate at CO_2 levels up to 8,000 ppm.[3] Although it had been believed that it's only well above 10,000 ppm that health issues arise for people, recent studies have shown incidents of headaches for astronauts on the International Space Station increase with rising CO_2. NASA's current standard for maximum allowable CO_2 in spacecraft is 7,000 ppm.[4]

But there's little information about optimal levels of CO_2 for plant growth or how much they're impaired by elevated levels. Most plants studied increase their rates of growth if there are moderate increases in CO_2 as long as other conditions are favorable: adequate nutrient and water availability and suitable temperatures. Greenhouses often supplement CO_2 to 1,000 ppm to increase growth since their crops are supplied with everything else they need. In our global environment, the effects of rising CO_2 and climate change on plant growth are far more uncertain.

A few plants, such as wheat and rice, have been tested at a wide range of CO_2 levels. While 1,200 ppm seems to be optimal for their seed harvest, doubling CO_2 levels to 2,500 ppm can reduce yields by one quarter. But higher levels (even to 20,000 ppm or 2 percent of the atmosphere) only reduces harvests by another 10 percent. Further complicating the issue, wheat and rice's vegetative growth increases with elevated CO_2.[5] But other crops behave differently. The vast majority of food crops have not been investigated, not to mention the hundreds of plant species in our wilderness biomes. So we were off the map of what was known—terra incognito!

Another urgent reason for concern is that ocean water absorbs CO_2. The more that's added, the more acidic the water becomes. This is critical for the health of tropical coral reefs, which normally live in waters with a pH of 8.2 to 8.4. This is on the basic side, rather than acidic, because pH 7.0 is neutral. Higher pH is crucial in enabling corals to deposit limestone (calcium carbonate) and grow new tissue.[6] Coral reefs are the biosphere's most prodigious builders. Coral reefs have deposited more limestone than all the buildings humans have constructed.

When Biosphere 2 closed, no one knew at what level of pH coral growth would be adversely affected, so we tried to limit the levels of CO_2 in the atmosphere. Gaie and her marine team used calcium and carbonate salts to buffer the ocean during the two years to limit the fall in ocean pH to safeguard our ocean's health.

The CO_2 Campaign Begins

The air was flushed from Biosphere 2 before closure, so we started at about the same CO_2 concentration as outside air: then roughly 350 ppm. We didn't expect it to stay so low, as we were entering the fall and winter months with shorter day lengths. Over the next month, CO_2 levels rose to above 1,500 ppm average.

In Biosphere 2 all the plants receive sunlight or enter nighttime at the same time. Every day there are two sharply contrasted phases. During sunlit hours, photosynthesis dominates, and CO_2 is removed from the atmosphere so that concentrations can drop 500 to 600 ppm. At night, respiration—the breathing of plants, animals, microbes—releases CO_2 so it can rise a comparable amount. Earth's biosphere is more balanced with half the planet in daytime and half in night. But the main reason CO_2 fluctuates so little is our atmosphere provides a large reservoir compared to plant photosynthesis or respiration by microbes, plants, and animals. The largest short-term fluctuation is when the northern hemisphere enters spring, and CO_2 drops around 5 ppm. In modern times, CO_2 has been slowly but steadily increasing. The rate of increase is accelerating; the Earth's atmospheric CO_2 now increases by more than 2 ppm each year.[7]

We formed a CO_2 team headed by John Allen, who closely liaised and coordinated with Gaie. Taber, Linda, Gaie, and I were the key operational team members inside. We met frequently to review the situation, plan strategies, and schedule work crews.

Every breakfast meeting, I brought down my notebook to read off the previous day's high and low CO_2 measurements along with weather forecasts. We measured our challenge in ppms and moles (a quantity of light). In southern Arizona, the summer maximum day length is 14.5 hours; in dead of winter it falls to 9.5 hours. The yearly maximum is 70 moles per day of sunlight in the summer. That drops to 44 moles at the beginning of fall and bottoms out at 27 moles on a full-sun day at the start of winter. But inside Biosphere 2, we received only 40 to 45 percent of the outside sunlight.

Cloud cover decreases moles of light even more. We quickly learned Biosphere 2 was highly solar-powered. Our automatic CO_2 sensors recorded every fifteen minutes in our rainforest, savanna, desert, and farm. We could tell a cloud had come between the sun and Biosphere 2 by looking at a computer screen in our command room (which housed our desks and banks of large system computers). There would be a small decline in the rate CO_2 was being drawn down.

The terrible news we could not ignore was that CO_2 was not holding steady—even in the early weeks of our first fall!

A storm front bringing cloudy weather could increase CO_2 levels hundreds of ppm per day. It seemed cruel luck our first year in Biosphere 2 was an El Niño one. That weather pattern shifts the jet stream south from the northwest United States. It means higher rain and more cloudy days in the southwest, where we were. These El Niño weather conditions lowered average sunlight by 10 to 20 percent through our fall and winter months. We already had an extra handicap our crucial first months inside.

CO_2 was rising rapidly during moderate October day lengths. What could we expect with the shorter days of late fall and early winter? The New York *Village Voice* had headlined a prediction that we'd have to leave Biosphere 2 by our first Christmas, just three months after entering, because of the increase in CO_2. One of our computer analysts in mission control was so confident, he made a similar and public bet.

We were determined to prove them wrong. CO_2 in Biosphere 2 was not a spectator sport—we intended to work intensively to make sure it didn't get out of control.

Green Allies and "Sunfall"

Our main allies were the green plants. The faster they grew, and the more of them there were, the more CO_2 they'd pull out of the atmosphere. Plants and trees are also our allies in the global biosphere. Planting a tree does more than provide greenery, shade, timber, and fruit. On average, one tree stores forty-eight pounds of CO_2 per year, one ton over forty years of growth. One large tree also provides two people with oxygen and improves air quality. Trees around buildings can lessen air conditioning and heating needs by 30 percent or more, helping reduce the amount of greenhouse gases released.[8]

We began our two-year meditation on "sunfall." Sunlight was the main limiting factor for plant growth. Receiving less than half of outside light because of space frame glass, structural shading, and the seasonal shortening of day length meant our first season was going to be dramatic. Other resources were abundant enough. We recycled our water and could supply as much as plants needed in the farm and biomes. Our soils were fairly rich, so the plants had enough nutrients. Sunfall couldn't go unharvested if we were to win the CO_2 battle—any place sunfall wasn't intercepted by a green plant wasted one of Biosphere 2's most precious resources.

Frequent CO_2 "blitzes" included getting plants into bare areas. We planted sugarcane on the cloud forest atop the rainforest mountain; we potted plants in containers along the savanna cliff-face wall; and we set up hundreds of planters in the agriculture basement promenade where they could get

Figure 17. Part of our research on carbon dioxide dynamics involved the apparatus we dubbed "R2D2." It measured how much CO_2 came out of the soil. We used it to study our biomes, especially in the transitions between their inactive and active seasons and our farm soils after soil disturbance for planting. We correlated this data with soil moisture content. I am extracting a sample in the savanna next to R2D2 with a soil corer (photo by Abigail Alling).

sunshine. Spare lights were deployed above a bed of fast-growing algae and vegetables in the farm basement.

Test module experiments taught us what soil disturbance did. During Linda's twenty-one-day experiment, we noticed a small sharp spike in CO_2 in the middle of a sunny afternoon. Linda solved the mystery. She had harvested a sweet potato from the vegetable patch. Digging into the soil with a hand trowel released a burst of CO_2. Soils contain five to ten times more CO_2 than the air.

This presented a new problem: we couldn't stop turning soil in our farm plots. Quick turnover after harvest started another crop growing, providing valuable food. Getting the crop in the ground quickly was also good for CO_2 management, as most crops slow their CO_2 uptake near harvest. But we planted immediately after the soil was turned and did experiments with crops planted without soil tillage.

In addition, we limited other activities that released CO_2. We stopped composting by mid-autumn, since CO_2 is released as the compost pile heats up and starts decomposing. Worm boxes were shut down for the same reason. They'd be reactivated when day lengths increased and we were past the "heart of darkness," the dreaded term we named our first winter. We celebrated the beginning of winter with a solstice "Photon Feast," toasting every time the sun came out from behind clouds. It was a hopeful day; we reminded ourselves that every day after December 21, we'd get a few more minutes of daylight.

Carbon Storage (Sequestration)

Teams pruned plants capable of rapid regrowth. These included the ginger belt perimeter plants, which wrapped around the rainforest, reeds, cattails in our freshwater marsh and grasses in the lower and upper savanna. The desert, thorn scrub, and savanna were our "biovalves," since depending on the season, they were active or dormant. Within the limits of their health, these biomes could be activated a bit early and kept growing a bit longer to help lower CO_2. The savanna was particularly important; its grasses grew lushly and were more tolerant of being active longer. So we cut the grasses, some capable of five- to six-foot growth a year, so they'd be ready for major carbon uptake when we turned on the rain.

Left in place, these cut-plant materials would decompose rapidly and release the stored carbon as CO_2. We bundled and dragged the pruned biomass to drier places, like the basement or lung, to slow decomposition. Normally nature takes control of carbon sequestration or storage. But we biospherians were growth-accelerators and carbon storage helpers. We hauled tons of pruned biomass. Part of the job description: matter-moving machines.

In our global biosphere, carbon sequestering is done on a grand scale. Limestone (calcium carbonate, or $CaCO_3$) is deposited in marine environments, and ocean water absorbs CO_2. Terrestrial vegetation and soils function as large carbon sinks; plants use CO_2 to grow tissue and store more carbon than they respire. Though true of cropland, rangeland, and land biomes, forests, with their large standing biomass, store the most carbon. Two and a half acres of forest absorbs three tons of carbon per year.[9] That's why deforestation contributes to greenhouse gas increase. The clearing and burning of forest lands releases carbon stored in the wood. Plus, the land's conversion to crop or pasture results in a net loss since those store less carbon than the original forest trees.

Methane (CH_4) frozen into the tundra soils of polar regions is a major global carbon storage. It is not yet known when global warming will begin thawing these permafrost areas that also extend underwater on continental shelves like those of Siberia. If permafrost thaws, it could result in an enormous and potentially catastrophic increase of methane in the Earth's atmosphere. Methane is thirty-five times more potent as a greenhouse gas than carbon dioxide on a one-hundred-year basis, and it is eighty-four times more potent over a twenty-year period. The additional methane would trigger further warming, thus leading to a positive feedback loop with more permafrost melt, methane release, warming, and so on. Thus far, there has been only a small spike in world methane levels since 2007, and the sources have not been definitively identified. In addition to permafrost melting, the release of methane may be coming from increased coal extraction and fracking (hydraulic fracturing) for natural gas production.[10]

During our first fall in Biosphere 2, we cut the savanna grasses and allowed it to "rain" (with our irrigation system), keeping the savanna growing through the fall and winter. We also "turned on" the desert and thorn scrub in stages through late October and November 1991, observing the plants for any signs of stress. Visitors to Biosphere 2 were treated to the sight of biospherians happily dancing in the rain.

Better Sequestering through Chemistry

An additional tool for our CO_2 management was our carbon dioxide scrubber. This unique and innovative system imitated how our global biosphere makes limestone, taking carbon out of circulation. Over geologic eras, marine organisms deposited huge quantities of limestone using calcium and CO_2 from ocean water. Our chemical precipitator employed a several-step chemical reaction to make powdery limestone.

The beauty of this system was that it was reversible. When Biosphere 2 needed more carbon dioxide, the reactions could be reversed. Then by heating the limestone, the liberated CO_2 would go back into the atmosphere. We joked it would then be Biosphere 2's volcano. Volcanoes release vast amounts of CO_2 when they explode, putting carbon previously tied up in mineral- and life-deposited forms back into circulation.

The steps in our chemical precipitator that took CO_2 out of the atmosphere are listed below. Heating in a furnace at 950 degrees Celsius would release CO_2 back into the atmosphere.[11]

1. $CO_2 + 2NaOH \rightarrow Na_2CO_3 + H_2O$
2. $Na_2CO_3 + CaO + H_2O \rightarrow CaCO_3 + 2\,NaOH$
3. $CaCO_3$ (+ heat) $\rightarrow CaO + CO_2$

When would Biosphere 2 be short of CO_2 and need its release? Time would tell, but when the plants and trees inside grew to their full sizes, the fifteen tons of living biomass we started with might increase to seventy tons, almost five times as much. By that time, the soils, originally enriched with compost to ensure there were enough nutrients to power food crops and rapid growth in our biomes, would also have matured and become more stable. In short, Biosphere 2 was an experiment and a balancing act—between abundant nutrients and organic materials (which contain carbon) for early development and the provision of resources later when Biosphere 2 grew up.

Our carbon precipitator absorbed about 100 ppm per day of CO_2. But it couldn't run limitlessly because of the amount of chemicals inside the biosphere. Predictions we wouldn't last the first winter included robust use of the device. The life systems of Biosphere 2 would be needed to prevent a runaway rise in CO_2.

Unexpected Victory

Almost to our disbelief, all this effort paid off. Each of our efforts may have only incrementally improved Biosphere 2's capacity to take CO_2 out of the atmosphere and reduced CO_2 output. But it all added up to a surprising larger victory. We turned off the carbon precipitator device by the end of the

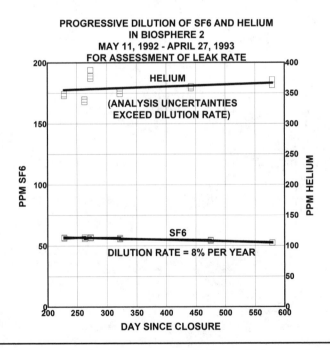

Figure 18. Two methods of determining the leak rate (tightness of sealing) in Biosphere 2 by the loss over time of the inert gases helium and sulfur hexafluoride (SF_6). W. F. Dempster, "Methods for Measurement and Control of Leakage in CELSS and Their Application in the Biosphere 2 Facility," *Advances in Space Research* 14, no. 11 (1994): 331–335.

first week in January 1992. On only partly sunny days during that month, Biosphere 2 either stayed even or dropped atmospheric CO_2 concentrations. We had risen to a daily high of about 4,200 ppm CO_2 that first winter. Then CO_2 began its long descent to the lows of our first summer.

We implemented these CO_2 strategies throughout the two-year closure. We kept close track of CO_2 levels and directions in movement daily. It was a major factor in operational decisions. We let the savanna, thorn scrub, and desert have their dormant seasons each year during periods when we knew we could control CO_2 rise. We reactivated compost piles and worm beds during the lower CO_2 months of spring, summer, and early fall. Each spring, some compost was delivered to the rainforest, savanna, and marsh to make up for the dried biomass we had removed from those areas in the fall. We had our own informal carbon trading market!

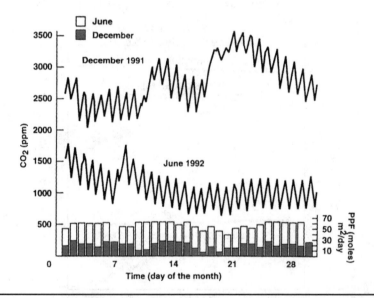

Figure 19. Biosphere 2's CO_2 levels were highly correlated with seasonal light. At the bottom is the sunlight for December 1991 (in dark boxes) and June 1992 (in white boxes). In December 1991, CO_2 levels were between 2,100 and 3,700 ppm, while in June 1992, CO_2 fluctuated between 800 and 1,700 ppm. The decline in late December 1991, three months into the experiment, meant there would be no runaway rise in CO_2. M. Nelson et al., "Using a Closed Ecological System to Study Earth's Biosphere: Initial Results from Biosphere 2," *BioScience* 43, no. 4 (1993): 225–236.

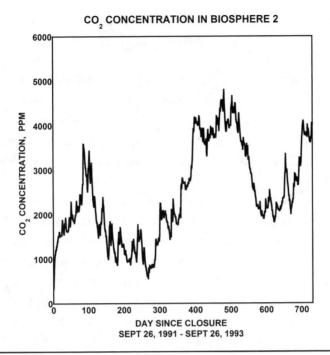

Figure 20. CO_2 dynamics in Biosphere 2. The two-year fluctuations of CO_2 included two peak periods during the two winters declining to lower levels during summer longer day months. M. Nelson et al., "Bioengineering of Closed Ecological Systems for Ecological Research, Space Life Support and the Science of Biospherics," in *Environmental Biotechnology Handbook* 10, ed. L. K. Wang, V. Ivanov, and J. H. Tay (Totowa, NJ: The Humana Press, Inc., 2010).

Linda termed our management of our atmosphere and carbon resources a "molecule economy." Cutting and drying out plant material and storing it in basement areas was like creating a carbon bank. "[We would] deposit the carbon into our account for safekeeping so that we can spend it next summer when we will need it for long days of plant growth."[12]

Our Global Greenhouse Gas Challenge

Our concern and management of CO_2 was not linked to worries about the impacts of global warming on Earth's biosphere: extreme weather events like hurricanes, drought and flood, rising sea levels, and the effect of higher temperatures on plant and animal communities. But there are lessons to be learned from our experience.

First, we humans have to get off the sidelines; our individual actions make a difference. Every incremental change in a positive direction is better than inaction or detrimental actions. We are also in a new relationship to our global biosphere—wilderness in our times is not the wilderness of preindustrial societies. Our cities, industry, and farms have an impact on the remotest areas of planet Earth.

For instance, recent data and a NASA visualization show that air pollution from Chinese cities travels around the world within months and makes clouds heavier. This increases the strength of storms and produces extreme weather events in distant countries.[13] Industrial activities and burning of fossil fuels increases atmospheric acidification, making acid rain no longer purely local or regional, but something that occurs around the world.[14]

We can choose ways of working with nature to prevent CO_2 (and other greenhouse gases) from rising and to mitigate its impacts. Greening our cities, suburbs, and industrial areas increases the number of green allies combating global warming. The sad truth is that the world is currently heading in the opposite direction. It is imperative that we stop deforestation. Thirty million acres of forest per year are cut down. Partly offset by the spread of forests elsewhere, mainly into abandoned agricultural land, the net loss is about fifteen to twenty million acres.

In just fifteen years, from 1990 to 2005, our biosphere lost 3 percent of its total forest area and currently loses seventy-eight square miles per day.[15] Deforestation releases 12 to 20 percent of total greenhouse gases.[16] In addition, this loss of forest further decreases the biosphere's capacity to remove CO_2 from the atmosphere. Two and a half acres of forest containing five hundred tons of carbon would be worth between $17,500 and $25,000 if carbon were valued at $35 or $50 per ton. That would make the forest far more valuable than its current monetary value of clearing it for wood and using the land for agriculture.[17]

Using more natural ways of growing our food crops has multiple benefits: food with less pesticide, less energy consumption, less CO_2 released.

Chemical fertilizers pollute our water and require large amounts of energy to produce and transport. Changing our world diet to less beef and milk would substantially reduce greenhouse gas emissions and the conversion of biomes to agricultural use.

Rebuilding carbon-rich soils stores enormous amounts of carbon. Healthy soils reduce the danger of soil erosion and reverse declines in fertility. One new inch of organically rich topsoil on the world's agricultural and range-lands would keep CO_2 from rising even further,[18] an approach called "regenerative organic" rather than sustainable since it heals and improves soils.

Global climate change results from nearly everything we do. Solving it requires many incremental and, if we can, major steps, like the many fronts pursued in Biosphere 2, to keep rising CO_2 levels from destroying the world. It's not just cars and power plants contributing to this problem, but it is sobering to consider that, on average, using one gallon of gasoline while driving releases around twenty pounds of CO_2.[19]

By comparison, an average person breathes out a little more than two pounds of CO_2 per day.[20] Since our respiration oxidizes food that plants make by taking carbon out of the atmosphere, our breathing is not a net contributor to global warming. But because part of our diet comes from meat, eggs, and dairy that come from livestock that eat plants, the choices we make in our diet are very significant. The greenhouse gas emissions from livestock production, especially for beef, dwarf those released from growing plant crops.

Awareness of our greenhouse gas challenge is rapidly growing around the world. But even with this awareness, sometimes even scientists who study CO_2 need a jolt. When some newspapers learned that CO_2 inside Biosphere 2 had risen above 2,000 ppm, they reported there would soon be incapacitated biospherians.[21] These reports reached Dr. Bruce Bugbee, an eminent plant physiologist at Utah State University who researches ways of increasing crop production for NASA's space life support program. He did something he'd never done before. He took a CO_2 sensor home to his tightly insulated house. To his astonishment, hosting a dinner party on a cold night caused CO_2 levels to exceed 4,000 ppm! He had no idea this would happen. City dwellers likely have no idea either: city traffic doubles or triples CO_2 levels, and air inside office buildings can exceed 2,000 ppm.

Population, Poverty, and "De-Carbonizing" Development

Population growth is a sensitive subject because of religious beliefs, cultural attitudes, the situation in wealthier and poorer nations, and gender equality issues. But there is a very high correlation between population growth and CO_2 increase.

Climate scientists emphasize the unprecedented speed with which humans are changing the climate. A 10 ppm rise of CO_2 earlier in Earth's history took one thousand years; now, that same rise takes far less than ten years.[22] Per capita greenhouse gas emissions are higher in richer countries than in poorer ones. It's not just or acceptable for wealthy countries, which have achieved their living standards relying on fossil fuels, to declare that poorer countries should not be allowed to do the same.

We're challenged to reinvent our farming and technosphere so poverty can be ended without increasing greenhouse gases. Ending poverty and providing a better standard of life and more educational opportunities, especially for women, are linked with the demographic transition from high birth rates to much more modest ones. Eventually birth rates may fall so low that a decline in world population will occur, as is currently occurring in some wealthier countries.[23]

De-carbonization of development using ecologically sound technologies is important everywhere, but especially in poorer, developing nations. Rapidly increasing renewable energy use, energy conservation, and less reliance on fossil fuels has begun the process.[24] Population control through improved living standards makes these challenges far easier.

From our experience in Biosphere 2, where we worked so hard to prevent CO_2 and other problems from deteriorating our biosphere and our health, a big question still remains: Why do we play dice with our global biosphere? The place to experiment and learn about biospheres is in laboratories like Biosphere 2—not with something as precious and irreplaceable as the life of planet Earth.

6

Farming as If There Were a Tomorrow

Living Off the Land

WE WERE FIVE AMERICANS, two English, and one Belgian. None of us had come from a farming background. We weren't country kids. We were educated, middle-class city slickers, though a few of us had done some vegetable and fruit growing. Sally had been an avid gardener when younger and had worked on tropical development projects in India and in Puerto Rico. I had managed the vegetable garden and planted a fruit orchard at Synergia Ranch in New Mexico. We also grew vegetables and tropical fruit at Birdwood Downs in Australia, but we were far from self-sufficient. The stakes for all of us biospherians were about to be seriously raised. We were in a state-of-the-art, $150 million facility. And we were about to transform into subsistence farmers, living off what we grew.

It was a shock. We'd trained in and helped develop what became the Biosphere 2 agriculture system, growing crops both in research greenhouses and in seven weeklong simulation experiments inside the facility before full closure. We started the two years with harvested food from crops grown in Biosphere 2 before closure.

Ours was a half-acre farm. No wonder it was called the IAB, the Intensive Agriculture Biome. The main farm area had sixteen plots where we could rotate crops. The farm also had a line of banana and papaya trees and soil rice paddies. We grew other grains (wheat, oats, sorghum, and millet), starches (sweet and white potato, yam, and taro), beans, peanuts, and a wide variety of vegetable crops. Satellite sections included a small orchard with tropical fruits and a basement area near south-facing windows with rice paddies in fiberglass tanks, additional fruit trees, and our constructed wetland for wastewater treatment and recycling.[1]

I noted soon after we started the experiment:

> Providing eight people with a nutritionally adequate diet—a diet that includes milk, eggs, fish, meat, and a wide variety of plant food—on a half-acre of growing space has never been done before. Even the vegetarian diets of India or China need far more land per person . . . It wouldn't be easy—it would no doubt be close and dramatic, even without the risk of serious crop loss to insects or disease. And how many hours a day would we have to farm to grow all the food we need? Would we all become farmers with no time left over to do research in the rest of Biosphere 2? Kevin Kelly of *CoEvolution Quarterly* worried that we would become "eco-peasants!" There's a delicate irony in our challenge. Here we live in a facility with some of the most innovative technologies ever developed. Yet our fundamentals of survival unite us with almost every other human since Adam and Eve were metaphorically booted out of the Garden of Eden and forced to "dig in the earth" and work for their living.

Looking back, this must have been some wishful, or motivational, thinking on my part. Based on crop production in our research greenhouses and in Biosphere 2 before closure, it did not look good for producing all our food. And that was before we fully realized that we'd get less than half the sunlight because of structural shading! Our original estimate was that we would get 65 percent of outside light.

We decided to use soil rather than hydroponics, which we had tried early on in support greenhouses. Biosphere 2 would thus be more applicable to farming as it is done around the world. Our farm was intended to demonstrate an agricultural system with high productivity, little inputs, and no use of toxic chemicals. Soil makes recycling easier. Inedible crop parts and animal wastes were composted to make rich soil amendments. We wanted a resilient and long-lasting farming approach; the final factor was rich agricultural soils teeming with life. The microbial life of Biosphere 2's farm soils would consume problem-causing trace gases.[2]

There Is No "Away"

The Biosphere 2 farm went beyond organic standards in avoiding toxic chemicals. Some natural insecticides were proposed, like pyrethrum from chrysanthemums. Yes, we could use it to control pests, but then what? As Dr. Richard Harwood, a Biosphere 2 agricultural consultant and sustainable agriculture chair at Michigan State University, noted: "there is no away in Biosphere 2."[3]

You can't fool yourself into thinking you can truly throw anything away or that something just *goes* away. Where would it go? In such a small system, with a tiny atmosphere and a tiny amount of water, even minor pollutants made a difference. With the goal of a long lifetime of food production and research, small amounts might build up over years, even decades.

We limited ourselves to using soap sprays and BT (Bacillus Thuringensis) bacteria, which controls caterpillars. Our Integrated Pest Management (IPM) started by selecting crop varieties with better disease and pest resistance. We also used beneficial insects, like ladybugs, which feed on crop pests, nontoxic sprays, and even hand control (squishing the bad bugs).

Avoiding monocultures is vital. The impact of one poor harvest or unsuccessful crop is not devastating for overall food production if you're growing a wide mix of crops. We grew eighty crops, including herbs. There were at least two to three varieties for each crop, since each differed in resistance to disease or insects.[4]

Broad mites unexpectedly decimated plantings of soybeans, yard beans, and white potatoes. We switched to tropical cowpeas and lablabs for our beans and sweet potatoes (with some taro and yams) for our main carbohydrates.

Animal Farm

Subsistence farming really focuses your mind on the fine art and science of growing food. Biosphere 2 agriculture included domestic animals to provide some fat in our diet, which we worried might be too low. Animals were also part of our recycling program to maintain soil fertility. Our animals ate the human-inedible parts of our crops like stems and leaves. The manure of our domestic animals gave us nitrogen to heat up our compost. Though raising animals took time (around a tenth of total biospherian labor), interacting with the animals was also fun. Though we ate a mostly vegetarian diet, the small amounts of eggs, milk, and meat were made into special dishes that greatly enlivened our meals.

In our animal barnyard, we raised chickens, pigs, and goats. Pygmy animals made sense for our intensive but limited agriculture since we had to grow the fodder for our animals inside. There were no bags of store-bought animal feed. African pygmy goats were selected as our milkers, and Ossabaw pigs as a meat source. These pigs naturally dwarfed themselves living on small islands off the coast of the southeastern United States.

Most Americans and Europeans are divorced from what used to be a part of ordinary life for earlier generations. That we biospherians would kill and process our domestic animals caused disbelief and curiosity amongst the visitors. The animals wouldn't be sent off to a slaughterhouse and returned in plastic-wrapped packages. Most of the biospherians had experience in slaughtering at other Institute of Ecotechnics projects; a stun gun made slaughtering quick and painless for the animals.

Caring for the animals brought its pleasures. The goats and pigs all had names. Observations of their behavior frequently enlivened our conversations. The milking goats included Milky Way, Stardust, and Vision; the buck was Buffalo Bill. The adult pigs formed a cute family. Zazu, the female, larger than Quincy, the male, was the boss. She nipped him to get first pickings of their food and kitchen slops. They bedded down together in a wooden box each night. A few of the quirkier chickens got names as well, such as Mrs. Fruitcake, who seemed, for a chicken, to be a bit daffy.

The goats were delightful. The kids were irresistible, frisky, and playful. Quite acrobatic, they'd sometimes jump on their mothers' backs for a better view of the surrounding area and a free ride. The first live link to Good Morning America, the ABC early morning show, caused great debate amongst us and our media advisors. Should Jane, who along with Sally did the bulk of the milking and goat tending, have a goat kid on her lap during

Figure 21. Sally holding a baby goat. There were five born during the two years.

Figure 22. Jane feeding peanut greens to goats in the animal barnyard (photo by Abigail Alling).

the interview? That adorable and photogenic goat was the first born inside. But . . . what if they asked about that goat's future? We could only support four milking goats and a buck for breeding. The question didn't come up.

The animal pens were green and colorful with lots of ornamental and food crops growing in planting boxes. The plants sucked up CO_2 and made it pleasant for the domestic animals and us, their caretakers. The pigs could scratch and wallow in small water tubs. The goats, with a mountain ancestry, could climb wooden boxes.

Two chicken pens each had a resident rooster. Our chickens were a mix of scrappy Mexican jungle fowl (one of the ancient chicken lineages), elegant Japanese silky chickens, and a few Bantams (with their lineage's distinctively cocky attitude). We laughed watching competing roosters try to outdo one another with their *cock-a-doodle-doo*s and leaping high at each other on the wire mesh dividing fence.

Our rice growing emulated the rice-azolla-fish system of Southeast Asia. Tilapia fish tolerate nutrient-rich water and crowding, and they grow rapidly. The fish fed on azolla, a high-protein floating water plant, while azolla and fish wastes fed the rice. We grew rice seedlings in a fiberglass basement tank. Like Asian farmers, we'd plant the seedlings barefooted, enjoying the mucky mud. When a rice crop was harvested, we'd net the fish, sending the larger ones to the kitchen and transferring the smaller to other rice paddies. We had no surplus crops to feed the fish, so growth was slow. Fish was a rare dinner treat.

We harvested fast-growing, high-protein Leuceana trees, elephant grass, azolla from rice paddies, and foliage from constructed wetlands for fodder. Occasionally we'd feed the animals savanna grass prunings. After a year, we replaced the Leuceana with sweet potatoes, as they were far more productive; we ate the tubers, and the animals loved their plentiful vines.

One of my jobs in the farm was chief fodder collector. Part of my daily routine, I collected enough fodder on Fridays and Saturdays to get Sundays off. Each day we supplied eighty to one hundred pounds of green fodder to our animals.

Jane cooked up kitchen scraps daily, and we also fed the animals sweet potato or peanut greens and the stems of grains. Later in the closure, cockroaches became a problem. Sally made a shocking discovery: cockroaches climbed on and munched on some of our crops. She thought they found our

Figure 23. The author carrying fodder (sorghum stalks) after a grain harvest (photo by Abigail Alling).

horticultural oil sprays, used to combat mites, a "tasty salad dressing."[5] Four types of cockroaches in our wilderness biomes helped with recycling dead organic matter. They caused no problems. But the common house cockroach, a stowaway, exploded in numbers in the habitat and farm.

So an unusual duty was added to night watch rounds. Jane described our cockroach war:

> It did not help to know that roaches are an unsolved problem in many large botanic gardens . . . Although we kept everything spotlessly clean, once the kitchen was plunged into darkness, hordes of insects turned the white countertops brown . . . So the person on night watch had the chore of creeping into the kitchen to catch them unawares. Armed with a vacuum cleaner, he or she flipped on the light and vacuumed up as many roaches as possible before they all scuttled away. We then carried the roaches down to the animal bay and fed them to the chickens, which although startled awake, leaped into action chasing after the bugs, which were a great source of protein. Thus roaches were transformed into a sparse supply of chicken eggs . . . I gained a certain amount of perverse entertainment from the notion that the biospherian "heroes" had to fight insects at every turn for our place in this new world's order.[6]

Animal products supplied only a small portion of the Biosphere 2 diet. We got around forty ounces per day of sweet goat's milk. This made possible cheese, ice cream, fruit smoothies, and special sauces. We ate a small portion of meat (generally one quarter pound per person) once a week, and at special dinners, holidays, and birthday feasts. Later two of the crew elected to go entirely without meat, instead consuming more beans to compensate. The small amounts of meat weakened their ability to digest it, and they suffered badly from intestinal gas. Actually, intestinal gas and stomach pains after our large feasts were some of our major health issues!

Hunger and Health

We experienced various states of hunger throughout our two years. It wasn't unhealthy, as we were serendipitously (and by necessity) eating a very similar diet to the high-protein, low-calorie diet pioneered by Roy. This diet has led to great health results, from sharply lowered cholesterol levels to enhanced immunity. In laboratory rats, it led to a healthy, exceptionally long life, a 50 percent increase over those eating as much as they want. Roy called his popular book *The 120-Year Diet*.[7]

We were the first humans researched on the diet in a controlled environment where we couldn't cheat (sneak off to get more food) or lie about what we ate. We were lucky Roy was our doctor. He later said: "I think if there had been another nutritionist or physician, they would have freaked out and said, 'We're starving,' but I knew we were actually on a program of health enhancement."[8] He called it a "healthy starvation" diet, with lower calories but plenty of protein and nutrients from the fresh produce. Roy delightedly pursued the research opportunities the Biosphere 2 diet made possible. We willingly subjected ourselves to yet more medical examinations and physiological tests.

Every harvest was carefully weighed, and the amount of food doled out to the daily cook was recorded in Roy's nutritional database. We showed the same results as the lab animals, including reduced blood pressure. Our cholesterol dropped from an average of 190 to 120, levels usually seen only in children.[9] We laughed that our two years inside might pay off in extra years afterward. In a way, we were getting younger and healthier—another part of the time machine.

But no one, including Roy, was happy with the reality of ever-present hunger. The first three months inside, we averaged 1,800 calories a day though we worked six days a week, with everyone doing three to four hours of farming on weekdays.

Subsistence farming was a shock to digest. There was no running out to a convenience store or fast-food joint, no in-between meal snacks or late-night trips to the fridge. We summarized this state of affairs: "If we don't grow it, we don't eat it!" Our black pepper plants never flowered during the two years, so we went without. Our sardonic comment about pizzas produced for special feasts was that they were "better than the competition for sure. But delivery time is a problem." As Sally noted:

> Our favorite dish, biospherian pizza, took at least four months. This was the time it took for the wheat crop to mature, not to mention the time to thresh and grind it. Then there were tomatoes and peppers and onions to go on

top and goat cheese made from milk. (We needn't go into the time it took to grow fodder to feed the goats to produce the milk, etc.! The story could go on forever.)[10]

Linda thought some of the problem at the start was that "we just plain were not good farmers." But we eventually got better—hunger is a great motivator. Early harvests of sweet potatoes were long on green vines and low in tubers because our rich soil promoted strong vegetative growth. We learned to dry-shock (turn off the irrigation for spells) and heavily prune the vines, which signaled the roots to get busy making tubers.

We researched optimal environmental conditions for our forty major crops. We learned the best time of year and temperatures for our temperate crops (wheat, beets, greens) and semitropical crops (sorghum, lablab, millet, sweet potato, peanut). Some crops we only grew in low-light seasons and others we grew in long daylight seasons. Papayas in the low-light basement and short daylight winter months took a long time to turn orange and ripen, so we started picking them green as a vegetable. Though bland, green papaya makes an excellent base for porridges and soups. Picking them green led to more fruit production as well.

Food shortages meant we had a painful decision whether to keep the pigs. Originally, the plan was to feed them with extra starches, like sweet potatoes, to boost their diet. But with the low light of El Niño, there weren't any extra starches. After a year, even the adult pigs were slaughtered. We toasted their memory at feasts they enlivened.

Victory Gardens

The sunfall meditation led to our victory gardens. Originally, these gardens were a program of small food gardens ordinary Americans started when many farmers went off to the army during wars. In 1944, twenty million gardens produced eight million tons of food, supplying more than 40 percent of the nation's fruits and vegetables.[11]

Finding new places to grow food became one of my passions (obsessions may be more accurate). I managed the basement agriculture and carefully

Figure 24. Victory gardens in Biosphere 2. Another wheelbarrow of soil is headed for where plants could use unused sunfall (photo by Abigail Alling).

surveyed every square foot of ground receiving enough sunlight to support crops. By the time I was done, I'd installed more than one thousand extra pots and small planters made from spare materials. Laser and I teamed up— he scrounged spare lights, and we made planters in sunless empty basement areas.

The balcony was one of our favorite places to "eat out." Psychologically we felt "inside" while in the human habitat, so on most Friday nights and some holidays, we could be found at our Café Visionnaire overlooking the farm, whose balcony received some of the best sunlight. When the structural engineers confirmed it could support the extra weight of soils, we DIY-ed planting boxes to grow white and sweet potatoes and beets there. I hauled unproductive, shaded tree boxes from the basement to grow more bananas and papayas on the balcony as well. On some insomniac nights, I'd find myself doing this at three or four in the morning!

One highlight of the sunfall victory garden campaign was while I stood on the balcony with Laser, our master technosphere maintainer/improver, both of us looking out at our beautiful, life-supporting farm, contemplating how we could grow more food. I turned to Laser and said, "Stairways." We both saw it. There were four farm stairways down to the basement. We'd need the ones on the north. They were shaded by a line of bananas and papayas, and we needed them to haul harvests and equipment up and down. But we could cover the south and part of the middle stairwell with a plywood-bottomed planter and gain another 250 square feet of growing area.

We harvested more than two thousand extra pounds from victory garden additions our second year inside. With our farming learning curve, our calories increased ten percent from 2,100 kcal a day for the first year to 2,300 kcal for the second.[12] The body gets more efficient at extracting nutrients from food when you're on the calorie-restricted diet, so the crew regained a bit of the large weight losses we'd had the first year. Over the two-year closure experiment, the women averaged 10 percent weight loss, and the men 18 percent.

Though we fell short of replacing all our food, we did supply 83 percent of our diet from what we grew during the two years. Richard Harwood, our agriculture consultant and an expert in traditional Asian agriculture, calculated our half-acre farm was the most productive in the world, exceeding by "more than five times that of the most efficient agrarian communities of Indonesia, southern China, and Bangladesh."[13]

Growing food was the crucial first step, but we had to do all the post-harvest processing. We had threshing machines for grains and beans, drying ovens, sugar cane presses, drying racks for potatoes, fruit, and vegetables. For the sheer fun of it, crew members laid dried bean pods on a tarp and did a wild dance to break them open. Sally had her own method:

> Beans could have their hulls removed without the machinery. I discovered the best procedure was to put them in a sack after they'd been dried then hit them for about five minutes with a rubber mallet. After that they could easily be put through the winnowing machine, which separated the broken hulls from the beans. It was a very therapeutic activity; one could visualize that the sack was one's worst enemy and hammer away. I always recommend it to anyone who's feeling angry or aggressive.[14]

The Passionate Farmers

Looking at each meal and knowing how every component of every dish got there, from seeding to weeding to watering to harvesting to processing, was extraordinarily gratifying. We were purists: we grew all the herbs and spices we used and even collected salt for the table from space frames overlooking the ocean. Our only supplement was vitamin D tablets.

We grew passionate about our farming, motivated to lessen hunger and to see if we could reach 100 percent production. We didn't achieve the latter, having to dip into seed stocks our last six months inside. But the second crew of biospherians grew all their food during their six months in 1994. They benefitted from improvements we made during the transition, such as selection of more shade-tolerant crop varieties.[15]

How our crops were doing, strategizing on ways to combat insects or diseases, appreciation of the simple pleasures of working the soil, and caring for our plants filled a lot of our days and thoughts. A friend in England complained when I emailed him portions of my Biosphere 2 journal. Its endless

Table 4. Food production from different crops and domestic animals for the nutrition of the eight-person biospherian crew during the first two-year closure experiment in Biosphere 2, 1991–1993.

Crop	Total 2-Year Yield Kg	Grams Per Person Per Day	Protein (G.)/ Person	Fat (G.)/ Person	Kcal/ Person
Grains					
Rice	277	47	4	0.9	168
Sorghum	190	32	4	0.6	107
Wheat	192	32	4	0.7	108
Starchy Vegetables					
Potato	240	41	1	0.6	31
Sweet Potato	2,765	468	7	1.3	494
Malanga, Yam	2	20	12	0	22
Legumes					
Peanut, Soy Bean, Lablab, Pea, Soybean, Pinto Bean	208	60	13	13.2	269
Vegetables					
Beet Greens, Sweet Potato Greens, Chard	637	108	1	0.2	22
Beet Roots	760	129	2	0.4	57
Bell Pepper, Green Beans, Chili, Cucumber, Kale, Bok Choi, Pea	331	57	1	1	15
Carrots	225	38	0	0.1	17

Cabbage	153	26	0	0	6
Eggplant	245	41	0	0	11
Lettuce, Onion	289	49	0	0.1	11
Summer Squash	513	87	1	0.1	17
Tomato	353	60	1	0.1	12
Winter Squash	343	58	1	0.2	37
Fruits					
Banana	2,171	367	2	10.5	220
Papaya	1,216	206	1	0.2	53
Fig, Guava, Kumquat, Lemon, Lime, Orange	133	23	0	0.1	11
Animal Products					
Goat Milk	842	142	5	5.6	99
Goat, Pork, Fish, Chicken Meat, Eggs	94	108	3	3.1	38
Total Produced	12,432	2,107	53	39	1,823

Sources: S. Silverstone and M. Nelson, "Food Production and Nutrition in Biosphere 2: Results from the First Mission, September 1991 to September 1993, *Advances in Space Research* 18, no. 4/5 (1996): 49–61. ; M. Nelson, W. Dempster, and J. Allen, " Key Ecological Challenges for Closed Systems Facilities," *Advances in Space Research* 52, no. 1 (July 2013): 86–96.

Figure 25. Harvesting grain in the intensive agriculture biome. From left: Laser, author, Taber, Sally.

talk of crops, pests, and food reminded him of boring accounts of the farming struggles of early American colonists.

There is no doubt food became our obsession. Linda had friends fax in menus from restaurants in Tucson so she could vicariously enjoy the variety of choices. Jane admitted that though her English sensibilities and niceties were riled, she also became obsessed:

> We quickly invented the "biospherian serving," heaping as much porridge or soup into a bowl (the cook portioned most other food) . . . Not wanting to waste any food at all, we all took to sucking our chopsticks clean and wiping our plates with our fingers. This habit soon turned into blatantly picking up the plate and licking every morsel from it. It was a great insult to the cook if all the plates were not licked entirely clean. . . . We are turning into a group of barbarians! I lamented, but I licked my plate along with everyone else.[16]

An enjoyable pastime was to fantasize about our favorite foods, going around the table, each of us going into a reverie while we conjured past and future pleasures. Jane recalls:

After dinner some of us sat around . . . and engaged in a recurring form of therapy—food fantasies. We imagined and described in exquisite detail a rapturous meal we wished we were eating. Sometimes I could smell a flourless chocolate torte as I brought the empty fork to my mouth. As I placed it on my tongue I could feel the gooey, creamy consistency, and taste the full, rich, pungent, dark chocolate, and I washed it down with an imaginary cappuccino . . . We were hungry for stimuli. When watching a film, I would focus on the eating scenes, forgetting the plot. Taber and I would talk about what was on movie plates and in movie glasses. We took to watching cooking shows on TV.[17]

I wrote poems to our crops and domestic animals during the two years. Here's one.

Spud Service

I worship the potato
Building altars to its starchiness
Revering its stick-to-the-stomach essence
I pack my basement with images of my lord
I count the days until I harvest
The sun shines for my tubers
I water and withhold according to their needs
I'm torn between white and sweet
Fearing disloyalty will jeopardize the crops
I dread its enemies—my enemies
White I will nurture
Sweet I will dry shock
If I fail it's all fodder and compost
And I will seek taro or yam to adore

With hunger sometimes even immediately after leaving the dinner table, our cooking skills ramped up several notches. Each meal was important for our morale. The eight of us cooked once every eight days, starting with dinner, then the next day's breakfast and lunch. Even with dozens of crops, presenting food in new combinations and recipes was crucial. The vegetables from our productive patch added a lot of menu variety.

Sally, in charge of our food systems, wrote a Biosphere 2 cookbook naturally titled *Eating In*. There she recounts the difficulties we had:

> Some of the early attempts were awful. Our coarse home-ground whole-grain flours did not behave in the same way as store-bought flour; our meat was tougher than the cuts from the supermarket; and few of us had experience with our staple, the sweet potato, except to bake them for Thanksgiving dinner once a year. Slowly a Biosphere 2 cuisine began to emerge as we each learned to experiment with the available food and find more creative ways of cooking."[18]

A few of our crops were champion producers, like sweet potatoes and beets. We each ate about a pound a day of sweet potatoes after we gave up trying to grow white potatoes because of the dreaded broad mite. The last season of our closure, all the beta-carotene from sweet potatoes and beets gave our faces and skin a somewhat weird orange tint. Sweet potatoes and bananas were some of our sweeter foods and were key ingredients in many Biosphere 2 ice creams and cakes. Sweet potatoes served as the base of soups, porridges, and our oilless oven-baked fries.

Figure 26. Peanut farmers. Sally, Jane, and Linda share a laugh while processing a peanut harvest. Behind them is a crop of sorghum (photo by Abigail Alling).

We couldn't do normal frying; we decided to enjoy our peanuts roasted rather than making cooking oil. This inspired improvisations to many recipes. Beets were a greater challenge, though we grew golden beets, which had a somewhat different taste than red beets. After the mountain of beets we ate, some crew members swore they'd never eat beets the rest of their lives!

Early on, we decided that even though we were limited in calories, we'd reserve a portion of harvests, eggs, meat, and milk for feasts. We prepared special dishes for each feast and decorated our feasting location. We occasionally ate in different venues, such as the beach or our tower library. The feasts featured special foods, like goat cheese, sausage, and home-brew. These were banana wine (the best), rice beer (mediocre but still enjoyable), and the far less successful beet "whiskey," which everyone abhorred even while drinking it. Sitting down before a heavily laden table was pure heaven.

Another gustatory delight was the precious coffee made from beans from our young, small rainforest coffee trees. That happened only every two to three weeks on Sundays, but from the time the smell wafted down the hallways outside our rooms, spirits rose. I made lousy-tasting yerba mate, a caffeine-rich plant that grew in our rainforest. With no firsthand experience of the prized caffeine drink of South America, I finally gave up.

Sally, at first, collected our best recipes as a resource for future crews. But she realized a cookbook of our creations would also be a guide to healthy eating. She noted:

> Gradually the crew began to take great pride in inventing a new dish or a new use for a particular ingredient . . . So I began hounding the other biospherians to take note of the quantities they used when cooking. This was sometimes a problem as the most inventive cooks would often just throw some things into the pot, see how it tasted, throw in something else, and when they were done they would be unable to say how they got there. Soon creating a "cookbook worthy" recipe became a matter of pride and the crew began to record their efforts.[19]

If there's one thing seven of us could agree on it's that seven of us became much better cooks inside Biosphere 2. Sally remembered:

In the winter we often had more leafy green vegetables than anything else, which were not too popular. There was one biospherian who insisted on making almost inedible sauces and soups out of pureed green leaves and little else. Eventually the person, who shall remain unnamed, was told that if this continued he would end up wearing his dingy green delicacies. The cooking took a turn for the better after that.[20]

That was the only time during the two years I threatened another crew member. Sadly, it was Roy. I had been warned even before closure that he was a great scientist and nutritionist but a lousy cook. I, and probably all of us, braced for his meals. But no one suggested he be taken off the cooking rotation—that would have meant more work for the rest of us.

Roy indignantly denied it when I suggested his meals were suspiciously light. But, like isolated crews in Antarctica—and hungry people everywhere—we too had a problem with food theft. The banana ripening storage room was the only locked door in Biosphere 2. Finally, even supplemental monkey chow for our galagos (small monkeylike animals in our wilderness biomes) was put in there—it was too tempting. A notorious incident was when Sally's stored piece of feast cake in our back kitchen fridge was not entirely stolen, but carefully cut in half, a heinous crime!

In months without a birthday or holiday, we invented feast occasions: peanut harvest fiesta, national sweet potato day, whatever we could dream up. Being so dependent on sunlight, we made special occasions of the solstices (longest and shortest days) and the equinoxes (equal daylight and nighttime).

For all of us, hunger was pretty much a new experience. Probably we were like humans have been for most of our time on Earth—sometimes hungry and dependent on successful crops and livestock raising (or hunting) for survival. No wonder our psychology and metabolism have a hard time dealing with food abundance and the plentitude of meat products, lots of sugar and fat, and processed food.

In earlier times, humans gorged when food was available to tide them through leaner times. In Biosphere 2, hunger certainly focused our minds and led to an alert state, as difficult as it was to deal with. Our food was probably amongst the healthiest, being completely uncontaminated by chemicals,

Figure 27. A biospherian beach party. From left: Sally, Linda, Jane, Taber, and the author (photo by Abigail Alling).

growing in rich organic soils, and freshly harvested and cooked. Since our diet was predominantly vegetarian, with grains, beans, root crops, vegetables, and fruit making up most of our portions, our plates were large and filled with food. Biosphere 2 food was free of processed sugar and oils, with infrequent small amounts of meat, eggs, or dairy.

Growing Good Food: Feeding "Locavores"

I continued growing food after I left Biosphere 2. Sally restarted the vegetable gardens at Synergia Ranch and managed the sustainable rainforest timber project in Puerto Rico. In 2008, she began managing organic agriculture projects and training in Bali, working with Gaie and Laser and the Biosphere Foundation. After some years remaining fallow, my colleagues and I restarted the Synergia Ranch organic vegetable gardens, which we turned into a commercial enterprise in 2010.

Farming three-eighths of an acre, we've sent lots of food to our community kitchen and about twelve to fourteen thousand pounds per year to our local farmers' markets, food banks, co-ops, stores, and restaurants, as well as to our CSA (Community Supported Agriculture). It's been fun extending my knowledge of food growing, intensively cultivating a small area, and producing a cornucopia. The organic fruit orchard I planted in the mid 1970s at Synergia Ranch produces twenty-eight thousand pounds from four acres (when we escape the spring freezes and have a full fruit crop).

Participating in local farmers' markets and having the pleasure of watching people taste homegrown fruits and vegetables is always a highlight of the season. Customers often have visceral reactions to good, clean, local produce. For many, it awakens childhood memories, a vivid reminder of what food is meant to taste like. We have fun marketing our especially delicious organic peaches as "orgasmic peaches," and it's surprising the reaction to our supersweet beets and tasty greens.

That experience has sharpened my appreciation for the importance of the quality of our food. It's only recent generations who are rightly apprehensive about what exactly is in food. Large-scale industrial farming has plenty of problems. Food in the United States, on average, travels many hundreds of miles (some estimates are 1,200 to 1,500 miles) from farm to customer. The average homemade meal in the United States contains food from five foreign countries.[21] Transporting food produces greenhouse gases and pollution. "Locavore" eating, or eating food that comes from a radius of one hundred miles or less, keeps your purchase dollars in local circulation, strengthening communities and farmers. Often small farms use less chemicals than large agribusiness operations.

Agricultural Diversity

Crop diversity preservation is important for world agriculture. Humans have been farming for at least twelve thousand years. Natural evolution and human selection and breeding developed myriad crops suited to every microenvironment and soil. There are four thousand varieties of potatoes in the Andes Mountains where they originated.[22] Hundreds of different varieties of corn, cabbage, radishes, peas, tomatoes, cucumbers, and other vegetables offered by U.S. commercial seed companies a century ago has dwindled to less than two dozen. In the 1800s, North America grew more than seven thousand varieties of apple trees. Now, less than one hundred remain.[23]

Genetic diversity provides disease resistance and adaptation to partic-ular growing environments. The "Green Revolution" increased agricultural production but promoted just a few varieties of modern hybrid seeds. Their crop varieties grew prodigiously with chemical fertilizers, pesticides, and irrigation. Their work and industrial agriculture's spread resulted in the loss of thousands of local varieties that grew better for farmers who couldn't afford expensive inputs.

Traditionally, India grew some thirty thousand varieties of rice; now only ten varieties make up 75 percent of the rice. In the United States, six culti-vars grow 70 percent of the nation's corn, and nine varieties produce half of the wheat.[24] Relying on a handful of crop varieties makes a new disease or insect attack devastating to food security. The potato famine in Ireland and the damage to U.S. corn crops in the 1980s are prime examples. Since 1900, more than 75 percent of worldwide crop varieties have been lost.[25]

The World Summit on Sustainable Development in 2002 decided urgent measures are needed to preserve these important crop varieties. Now, many regional, national, and international organizations work to preserve crop diversity.[26]

Industrial Farming and Fast Food's Impacts

Chemical pesticides, insecticides, and fertilizers as well as deep plowing amount to an undeclared war on natural soil life and fertility. In Australia, ten tons of topsoil is lost *annually* for every ton of grain that's produced. Soil erosion and degradation rarely makes headlines, but over the past forty years, 30 percent of the world's arable land has been made unproductive. Over half of that soil winds up in water bodies causing pollution and increas-ing the risk of flooding. It takes 150 years for nature to replenish an inch of topsoil. Yet U.S. farms are losing topsoil at a rate ten times greater (India and China at thirty to forty times greater) than the replenishment rate.[27]

In prosperous countries, the ills multiply. Fast-food outlets and processed food have created epidemics of ill health, catastrophic increases in childhood- and late-onset diabetes, and obesity. A chef joked in a radio interview I heard that they all know how to increase customer satisfaction—add more cheese and bacon! Our evolutionary history makes us suckers, unable to resist those satisfying carbs and fats.

Rats given the choice of unlimited food would gorge and gorge, achieving short-term satisfaction with greater illness and far shorter lifetimes than the ones fed the high-nutrient, low-calorie diet. When asked how far calories could be lowered, Roy sobered us by saying, "We know it's too low when the rats start to die."

People need to listen to their bodies; no one is advocating a food regime that is not delicious and satisfying. But we also must realize that our survival—natural selection—instincts may not serve our overall well-being. Evolution only requires us to live long enough to breed and raise children.

Our Biosphere 2 diet was extremely low in fat, worrisomely lower than standard U.S. dietary recommendations. We grew peanuts, though they weren't as productive as other crops, to increase our dietary fat. Even so, we averaged only forty grams of fat per day, with only 20 percent from animal products. Medical checks when we left Biosphere 2 showed that body fat among the women averaged 13 percent, and the men 8 percent, similar to healthy professional athletes. Overall, we were in excellent health.[28]

Diet and Climate Change, Water Resources

How we produce animal products, especially beef and milk, has enormous ecological consequences. Livestock production produces more greenhouse gases (18 to 50 percent) than the world's transportation system (13 percent).[29] Methane from cows and their manure is a far more potent greenhouse gas than CO_2.

CO_2 is also released when forests are cleared for grazing and farmland for soybeans and other grains for livestock. Water use is stunning; 1,850 gallons for a pound of beef, 380 gallons for a pound of cheese. In the United States, 30 percent of our total water use is related to meat consumption.[30]

A world population of one billion in 1800 has grown to seven billion, with nine billion projected by the mid-twenty-first century. The world could easily feed even our growing populations if our diets change to vegetarian or, at the very least, to a diet where meat and dairy are consumed in smaller quantities. Particularly among younger people, this shift is already underway.[31]

Since our Biosphere 2 diet was mainly grains, vegetables, and fruit, we could grow a high percentage of our food on half an acre. Seventy-five to three hundred pounds of grass and hay and six to seven pounds of grain in feedlots produce one pound of beef. Chicken and pork are far more efficient, with chicken at about 2:1 and pork at about 4:1.[32] Half an acre supplies a complete vegetarian diet for one person, but ten times more land is needed to supply a meat- and dairy-rich diet. Rice, wheat, and potatoes require two to six times less resources than chicken, eggs, dairy, or pork.[33]

Our global food system has a tragic balance: one billion people are malnourished, hungry, and not getting adequate nutrition, and one billion are grossly overweight. Both cause ill health. Present world agriculture is not efficient either. One-third of total food production, 1.3 billion tons of food, doesn't get from farms to consumers and is wasted every year.[34]

Small scale chicken and pig operations are major sources of livelihood with less environmental damage than raising cattle. More than a billion people derive their livelihoods from livestock production, so change will need to minimize human and economic disruptions.

A change will have to come because world agriculture and food systems are far out of balance. Let's hope the alterations are by intelligent choice rather than us hitting the wall of biospheric limits.

The Garden of Kitchen Delights

I often remember the pleasures of our Biosphere 2 farm. After about six months inside, I recorded in my journal the sensations of a walk through our garden of kitchen delights:

On my way to the basement where I tend the constructed wetland system and planting boxes with rice, fruit trees and herbs, I pass fields of waving wheat. The farm is a mosaic of different planted crops, some wheat is golden brown awaiting harvest, others green with seed heads filling, others still looking like an overgrown lawn. The *potagerie* plot has its potpourri of different vegetables from lettuces to chili and squash, sweet potato patches cover the ground with vines, sorghum fields with heavy-laden heads, and white potatoes at several stages of growth. I duck and weave my way to avoid running into banana stalks hanging pendulously over pathways, with slim papaya fruit ripening on trees in their shadows. To my right I wave to a group of visitors peering into the windows in front of the rice that is just emerging from the mud of the paddies alongside the arching stands of tropical lablab beans. On my return to the kitchen, I pass through the tropical orchard—a dark, green shadowy world where banana trees soar twenty feet towards the space frames, taros wave their giant fan-like leaves, delicate citrus, fig, and guava trees stake out a position of middling height, and even grapes and pineapples grow in the understory . . . it's rare for me to leave the IAB without having to brush off a

Figure 28. Fish-eye lens photo of the Biosphere 2 farm. Poles are to support bean crops (photo by Abigail Alling).

few ladybugs, or without catching sight of our introduced green anole lizards (cockroach and aphid-eaters), and with the visual and sensual feast of a walk in our garden of delights.[35]

Regenerative Farming: Giving Back as Much as You Take

What was unique about Biosphere 2's agriculture, and what does it teach us?

We were diligent in returning nutrients to the soil and maintaining soil fertility. We started with a rich organic soil, and we returned virtually all the nutrients back to that soil. Despite some losses to insects and disease, we achieved high productivity without use of chemical inputs—our farm did not pollute our water or air. We certainly faced challenges keeping our farm healthy. There was some salt increase in our main water reservoir and in a few farm plots. The Leuceana tree plot came close to salt levels that would have negatively affected production. That plot during the second closure experiment would have tests with extra application of compost and heavy watering to leach out soil salts more effectively. The rice paddy soils showed denitrification stress after being in near continuous use in waterlogged conditions.[36] Changing to upland rice grown without wetland conditions was an option for future closure experiments, so those soils could fully aerate and get healthier.

We built Biosphere 2 to find out problems and develop methods for ensuring long-term farm soil and water health.

7

Déjà Vu

The Water Recycles

Closing the Loop

OUR FARM WAS radical in its design for total recycling of nutrients and water. "Closing the loop" is a goal in bioregenerative life support systems for space. The approach is also gaining traction in global thinking about creating systems to sustain the health of our global biosphere.

We were the rain-makers—the ones who programmed computer controls for the farm irrigation, mainly sprinklers strategically placed around the plots. The water that drained through our four-foot soil beds, "leachate water," was collected in tanks in the basement. A similar irrigation system supplied the rain for our wilderness biomes, and we also collected their leachate water to recycle back as irrigation water for rainforest, savanna, thorn scrub, and desert.

Leachate water was mixed with two other kinds of water in our farm irrigation supply: condensate water and treated wastewater from the constructed wetlands.

Condensate water was collected around Biosphere 2. This water is very pure since it comes from evaporation and plant transpiration, so a portion of it supplied our drinking water—it was a pretty humid tropical world, so there was a lot of condensate water. The rest was pumped to the agricultural

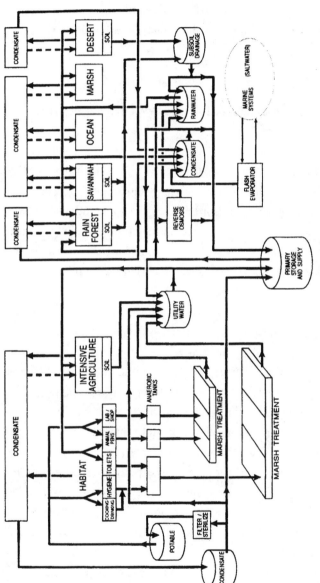

BIOSPHERE 2 FRESHWATER SYSTEMS

Figure 29. Engineering flow chart showing all the storages and flows in the Biosphere 2 recycling water systems. "Marsh Treatment" refers to the constructed wetlands for sewage treatment. M. Nelson, W. F. Dempster, and J. Allen, "The Water Cycle in Closed Ecological Systems: Perspectives from the Biosphere 2 and Laboratory Biosphere Systems," *Advances in Space Research* 44, no. 12 (2009): 1,404–1,412.

system storage water tanks and wilderness biome tanks to be mixed in with leachate water for irrigation. For both farm and wilderness biomes, condensate water helped reduce salt content in our irrigation. This was a major operational concern as we wanted to avoid long-term buildup of salts in our soils.

Constructed Wetlands for Wastewater Treatment

Treated wastewater from our constructed wetland system also fed back into our farm irrigation. Dr. Bill Wolverton, a pioneer in ecologically engineered systems based on nature, helped us develop the system.[1] Biosphere 2 had two separate constructed wetlands: one for human wastewater and one for animal liquid wastes and lab and workshop wastewater. Holding tanks collected all the wastewater, functioning like septic tanks, doing primary treatment by separating and digesting solids. Here, anaerobic bacteria did the work. Just like thoughtful homeowners with septic tanks, we were careful not to flush anything down our drains that could harm the bacteria. During the planning phase, we acquired soaps, shampoos, cleaning aids, and laundry detergents that were as natural and biodegradable as we could find.

Outside, a great deal of damage is done to septic tanks' living bacteria when people unthinkingly flush paints, solvents, waste oil, and nasty chemical cleaning aids down drains. The cost of chemical warfare against the microbes is more than ecological. Septic tanks don't do their job, and sludge builds up rapidly since the anaerobic bacteria normally consume and break down these solids. This necessitates expensive pumping out. In addition, with more solids and undecomposed organic materials coming out of the septic tank, leach drain soils get gummed up, preventing the wastewater from percolating. Leach fields fail far sooner than they should, and replacement is quite expensive.

Next we pumped the wastewater into our constructed wetlands. Normally, constructed wetlands are built in excavated soil with a waterproof liner preventing leakage before treatment is complete. In Biosphere 2, the wetland systems used a series of three fiberglass tanks. Recirculating pumps sent water back from the end to the initial tank.

Figure 30. Biosphere 2 sewage treatment system early in the closure. The plants grew rapidly, producing flowers, fodder, compost materials, and wetland habitat. Its treated water and remaining nutrients were used in farm irrigation.

More than a dozen types of wetland plants thrived in our constructed wetlands. They included floating plants like water hyacinth, azolla, and duckweed, as well as rooted wetland plants like canna lilies, bulrushes, and reeds.

Our constructed wetlands flourished. As manager of Biosphere 2's sewage system, I periodically cut the vegetation. The edible plants went for fodder, and the inedible, like water hyacinth, were composted. It was sewage treatment without chemicals and without a lot of machinery. The plants and the microbes in the water, soil, and plant root zones did all the work. The fodder increased milk, egg, and meat production. We harvested fifty pounds of animal fodder per week from our constructed wetland.[2] And I was proud. I reveled: "This New York City kid manages the sewage system for a whole world!"[3]

The constructed wetlands were also micro-ecosystems. Stowaway frogs found the wetlands to be a wonderful home, as did some ladybugs. I loved working in the wetlands, partly because I often did pruning and other

Figure 31. Cutting plants for fodder in the Biosphere 2 constructed wetland sewage treatment system. Pumps recirculated water between the tanks.

operations in full view of tour groups and visitors. I could see their aston-ishment when the tour guide explained this green and flowering garden was our sewage system.

An ultraviolet disinfection system at the end of the sewage treatment wetland was only used for research. We quickly learned after closure there were no infectious diseases amongst the eight of us. We "shared" our germs working together as a team prior to closure and were isolated from outside germs.

A green sewage system producing fodder, habitat, and flowers! But that wasn't all. The treated wastewater from the wetlands was sent to the farm

Figure 32. Just add wastewater, and it will grow! Some of the beautiful canna lily flowers in the constructed wetlands of Biosphere 2 (photo by Abigail Alling).

irrigation supply tanks. All the nutrients originally taken from the farm in the form of food we ate directly, or animal products from our fodder were returned to help maintain the fertility of our soil.

Rapid Cycling of Biosphere 2's Water World

We intuited water cycles much more rapidly in our small world than on Earth. Even in training, we biospherians reminded one another that what went into our water system anywhere would be in our tea in a matter of weeks. Later, these cycles were calculated. Water in Biosphere 2's atmosphere (humidity) only stayed there from one to four hours. That cycle is fifty to two hundred times faster than in Earth's biosphere. Our miniature ocean had a far longer average residence time, a bit more than three years. But this cycle is one thousand times faster than in Earth's vast oceans. Water draining through and evaporating from our soils had a similar cycling time as Earth's soils. But with a cycling time of less than two months, we were right to be super vigilant about not using anything that could pollute our soils.

The Present-Day Waste of Water and Nutrients

World agriculture, especially high-input industrial farming, makes little attempt to retain or recycle nutrients. Chemical fertilizers are used to boost production and make up for poor or degraded soils. Half those chemical fertilizers are washed away, polluting groundwater and surface water, without benefitting the intended crops. Crops and animal products are exported to towns and cities, but the nutrients don't return to the farms. Instead, energy-intensive, centralized sewage treatment plants receive the human waste, treat it like it's toxic, and disinfect it (sterilize it) generally with chlorine, a carcinogenic and environmentally harmful element.[4]

The almost universal final step dumps "treated" sewage into the nearest body of water—lake, river, or ocean. Treated sewage causes water eutrophication (too many nutrients) and fuels the growth of algae, which ties up oxygen, harming aquatic and marine life. Animal waste runoff also pollutes

Table 5. Size of water reservoirs and water cycling times in Biosphere 2 and Earth's biosphere. Soil water cycling is similar, since drainage through soils does not change much as long as soils are deep in a closed ecological system. In contrast, atmospheric water cycles fifty to two hundred times faster; in Biosphere 2's small ocean and marsh, water cycles one thousand times faster.

Reservoir	Volume (liters)	Percentage of Biosphere 2 water	Typical daily water flux (10^3 liters)	Estimated residence time—Biosphere 2	Estimated residence time—Earth biosphere	Acceleration in Biosphere 2
Ocean/Marsh	4×10^6	~61%	3.4 ± 0.4	~1,200 days	3,000–3,200 years	1,000 times faster
Soils	1–2×10^6	~23% (calculated on 1.5×10^6)	Irrigation: 18.3 ± 7.3 Soil Drainage: 4.6 ± 10.6 Plant uptake and ET	~60 days	30–60 days	similar
Atmosphere	2×10^3	~0.03%	12.4 ± 4.5 (evapotranspiration)	1–4 hours	9 days	50–200 times faster
Primary storage tank	0–8×10^5	~12% (when full)	$+4.6$ to 4.8 ± 3.1	~80 days		
Condensate and leachate tanks	1.6×10^5	~2% (when full)	$+2.8$ to 3.0 ± 9.7	~5 days		
Streams and pools in biomes	8×10^4	~1%				

Sources: F. N. Tubiello, J. W. Druitt, and B. D. V. Marino, "Dynamics of the Global Water Cycle of Biosphere 2," Ecological Engineering 13 (1999): 287–300. W. F. Dempster, "Biosphere 2: System Dynamics and Observations During the Initial Two-Year Closure Trial," paper presented, Society of Automotive Engineers Technical Paper No. 932290, Society of Automotive Engineers 23rd International Conference on Environmental Systems, Colorado Springs, CO, July 1993. W. F. Dempster, "Water Systems of Biosphere 2," (Oracle, AZ: Space Biospheres Ventures, 1992). M. Nelson, W. F. Dempster, and J. Allen, "The Water Cycle in Closed Ecological Systems: Perspectives from the Biosphere 2 and Laboratory Biosphere Systems," Advances in Space Research 44, no. 12 (2009): 1,404–1,412.

water, especially from feedlots and other high-density livestock operations. Those nutrients now causing problems elsewhere originated in our farm soils, which become even poorer for the loss of their fertility.

The consequences become horrific when the resulting nutrient soup—fertilizers, sewage discharge, animal and industrial wastewater—reaches the mouths of rivers and empties into marine zones. Now all the oxygen is tied up and dead zones are created, devoid of all marine life except for algae.

The first dead zone was discovered in the Gulf of Mexico, covering six thousand to seven thousand square miles. The same phenomenon has now been discovered all over the world. Dead zones are expected to double in size in the next few decades.[5]

In virtually all the poorer countries of the world, sewage is a huge human health disaster. Wastewater treatment is virtually nonexistent, so sewage contaminates drinking and household water. Eight hundred million people lack clean water, and 2.5 billion have no wastewater treatment. Waterborne diseases including cholera, typhoid, and dysentery spread. Diarrhea caused by contaminated water kills by dehydrating infants, children, and the infirm. Diarrhea kills three-quarters of a million children every year.[6] The World Health Organization estimates 1.8 million children die from waterborne diseases each year.[7]

Call This Progress?

So, there are two extremes. Richer countries spend vast amount of money and energy on sewage treatment and then throw away ("discharge") the treated wastewater. In the developing countries where sewage is not treated, there is widespread disease, death, and environmental degradation.

Contrast this state of affairs with practices that prevailed a century ago in cities like Shanghai. There, "night soil" (a wonderful term for human waste) was collected and delivered to boatmen who traveled up rivers to sell the material to eager farmers. The farmers composted the night soil and applied it to their fields. In this way, the nutrients were recycled and returned to the soils that originally grew the food sent to the cities. The cities made money from their sewage. A classic book written about traditional Asian agriculture called them "Farmers of Forty Centuries,"[8] a wonderful example of true sustainability.

Promising Alternatives

Is there a way to fix this mess? Modern sewage treatment plants face increasing requirements to remove more and more nutrients. If they can separate industrial waste (which includes heavy metals and synthetic chemicals that shouldn't go into soils) from normal residential waste, the resulting sludge, or bio-solids, would qualify as a modern version of night soil.

What about sending that back to all the farms and ranches supplying the city's food? The price we pay for food and sewage treatment should have an added cost to cover returning nutrients to our farm soils. Picture, perhaps, a system where the trucks delivering food to our supermarkets and food distribution centers receive an equivalent portion of composted sewage sludge to backload to the farms and ranches.

When correctly composted, human and animal waste pathogens are killed during the heating process when temperatures rise to 135 to 160 degrees Fahrenheit. Letting compost sit for six months is another safety factor against disease-causing bacteria.[9] The miracle of compost is that it makes organically rich, black soil out of wastes. Adding compost to the soil restores fertility, increases water retention, and soaks up greenhouse gases. Organic soils help decrease greenhouse gases rather than releasing them, as industrial farming does. Organic and more natural, traditional ways of farming don't rely on chemical fertilizers or pesticides to begin with, and humus-rich soils prevent loss of nutrients and soil degradation.

An increase in the number of locavores and a return to an emphasis on local farming in our diets would lower the costs of such a scheme. Some cities have begun composting their sewage sludge or organic materials separately from other garbage and solid waste. Others encourage homeowners to backyard compost or develop community composting of organic waste, kitchen scraps, grass clippings, and leaves.[10]

Hygienic and safe methods of using treated sewage for farming and gardening are being developed. These include constructed wetlands where the nutrients are utilized by green plants. Composting toilets save water in addition to enriching soils; ten tons of water flushes away each ton of human waste to our centralized sewage treatment plants.[11]

Gray water irrigation uses nontoilet wastewater where it is produced. This conserves potable water and naturally fertigates (water + nutrients) landscape plants. Droughts and water shortages are inspiring growing numbers of cities and regions to allow and even encourage these approaches with subsidies and rebates. In the face of necessity, changes can rapidly be implemented. Health regulations once strictly outlawed gray water reuse for irrigation.[12]

Taking Responsibility for Our Water Cycle

Taking responsibility for keeping our water clean in and out of the farm is key to improving human and biospheric health. Being a tightly sealed system, Biosphere 2 had an advantage in being able to recycle and complete the water cycle. We regarded our human habitat as a microcity. But the connection between what we did and what products we used in our microcity and the purity of our food and the health of its soils was clear and present. We began and ended the two years with the same water.

Early into closure, I noticed that on the weekends more wastewater arrived. The reason was easy to figure; on the weekends, the crew had more time for laundry and luxuriant showers. I loved that in Biosphere 2 there was no mystery in where water was coming from and where it went. We could literally follow the pipes and trace out our world's water cycle.

Salination: Destroyer of Civilizations

Agriculture uses 70 percent of the total human water supply and more than 35 percent of worldwide freshwater resources.[13] Farming use drives concerns about water shortages. Contaminants, including pesticides, chemical fertilizers, and antibiotics added in our agriculture and livestock operations get into soil and water resources. These add to the danger of salination of water and soils.

We had some problems with salt buildup in one of our farm plots and rice paddy areas. Salination was also a concern in the wilderness biomes—the main reservoir of Biosphere 2's water in one of the lungs increased its salt content during the two years.

Salination has majorly shaped human history since the invention of agriculture. Salt buildup from irrigation is blamed for the demise of many empires of the Fertile Crescent of the Middle East. The Sumerian civilization fell when rising sea levels and irrigation salt accumulation destroyed their soil's ability to grow food crops. The result was desertification and the creation of the present day "Infertile Crescent."[14] Books like *Dirt: The Erosion of Civilizations*[15] and *Collapse*[16] offer cautionary tales of how mighty cultures collapsed through salination, erosion, and desertification.

This challenge is still very much present. The UN Food and Agriculture Organization estimates there are 3.25 million square miles worldwide afflicted with salt buildup in every country and climate, but dry climates are especially prone to salt buildup, leading to desertification of once arable lands.[17]

Australia faces high risk, since 70 percent of its land is arid or semiarid. Salt buildup was unforeseen when native trees and plants were cleared for broad-acre farming. Water tables rose, bringing saltier water—the deep-rooted vegetation through their water uptake kept groundwater levels deeper. Seawater intrusion threatens food production in Australia's coastal lands where soil salt reservoirs have begun to affect the shallow roots of agricultural crops. Solutions being implemented include reforestation and restoration of native plants, better management of grazing and crop lands, and use of more salt-tolerant crops.[18]

In other regions, like the San Joaquin Valley in central California, a major breadbasket of the country, long-term heavy irrigation has increased soil salination and lowered water tables. Deeper extraction of water means using saltier water, which evaporates in the dry climate, leaving more salts behind to accumulate. Chemical fertilizers add to the problem. This situation, replicated worldwide, threatens more farmland loss and decreased water quality for urban populations.[19] Solutions will be costly and include soil improvement building organic matter content,[20] using natural fertilizers, water table management, and flushing salts with excess irrigation combined with methods of deep soil drainage.[21]

In one way or another, we all live "downstream" in our highly populated and industrialized planet. Everything "upstream" affects the water that all life depends on. Land pollution ultimately winds up in freshwater and oceans. Chemicals in those ecosystems are concentrated at higher levels of the food chain. The result is health concerns about fish and seafood harvested from polluted waters.

Even rain is not immune since modern technological activities impact water quality through air pollution. Rain is naturally slightly acidic but now has more CO_2 from industrial and farming practices, nitrous oxide from automobile exhaust, and sulfur dioxide from fossil fuels. Rain's pH has decreased from preindustrial levels of 5.6 to as low as 3.0 or 2.0, the level of strong acids. Acid rain can kill fish, damage forests, and harm cropland.[22]

The effects cause damage quite a distance from their source, since prevailing winds carry the pollution. Canadian forests are damaged from acid rain caused by power plants in the Midwest of the United States; Scandinavian forests suffer from industrial pollution that prevailing winds bring from England.

Even relatively clean rain becomes polluted when it contacts the ground on paved surfaces since it carries the residues of whatever contaminants are there: oil, toxic chemicals, trash, and pathogens. Storm water pollution is now recognized as a major cause of pollution of drinking water.[23]

Understanding What Goes Around, Comes Around

Inside Biosphere 2, it was a no-brainer to realize that our biosphere's health equaled our health. Everything we did anywhere inside would wind up in our air, water, and food. When I first met our Russian colleagues in the fields of closed ecological systems and bioregenerative space life support in the 1980s, I sent them boomerangs from Australia. It struck me that the most important feature of such tightly sealed systems was that "what goes around comes around." This phrase is equally true but much harder to grasp in our large, but far more tightly sealed, recycling planetary life support system.

Figure 33. View of the Biosphere 2 farm from high in the space frame. Along the left are the planting boxes on the IAB balcony (photo by C. Alan Morgan).

Just like the other cycles, the water cycles in Biosphere 2 were far faster than on Earth. This reinforced our palpable sense inside Biosphere 2 that every action counted. There's no hiding that reality when you're living inside a small biosphere. It made us think more than twice before we did anything that might add anything dicey to any water reservoir anywhere. Keeping our biosphere healthy became an act of self-preservation. Pollution of any kind was dangerous to us personally or by injuring part of the life support system. Even more, this understanding was far more than intellectual—it became deeply organic. Our bodies *got* it.

Eco-Farmer

I once sang of abstract fields of plenty
In folksongs of distant lands
Now I work in the garden of kitchen delights
Each day you perform the miracle
Of catching sunfall to harvest the air
While we tend you and groom you

Carry off your fruits
Deliver you the spent products
Of all you have wrought
Feeding your soil and watering your buds
Partners in keeping our world green
Put your ear to the earth
You will hear the cadences

8

Wilderness Biomes

The Biospheric Role of Wilderness

AS A MODEL BIOSPHERE, in addition to two anthropogenic (man-made) biomes, agriculture and microcity human habitat, this new world had "wilderness" biomes. They represented a spectrum of land and water biomes. Tropical systems were selected because they're the most productive, among the most threatened, and easiest to create in a southern Arizona greenhouse.

On land the gradient went from wettest to driest: rainforest, savanna, thorn scrub, desert. The desert occupied the low end of the wilderness wing and could tolerate higher summer and lower winter temperatures. The rainforest occupied higher ground on the other end. Our watery biomes included a marsh and coral reef ocean. The marsh features transition zones from freshwater inland marsh to coastal-fringing red mangroves. Our miniature ocean included beach, fore and back reef lagoon, and deeper open water areas.

The space life support community was mystified why we'd include areas analogous to wilderness. All previous life support facilities had limited themselves to food crops, algae useful in air and water regeneration, and living quarters for crew. We wanted wilderness ecosystems because, even for long-term space application, the beauty, diversity, and wonder of natural areas are important for human happiness and well-being. Who wants to live in a world with just buildings, food crops, and farm animals?

Figure 34. The longer wilderness wing of Biosphere 2. The Biospheric Research and Development Complex is in foreground, and the Mission Control building is across from the rainforest stepped pyramid on the right.

As a global ecology laboratory, we could investigate the impacts of farm and human habitat on wilderness. We hoped the wilderness areas would act as a safety factor, improving the air and water quality of the entire system. Gaining knowledge about the interplay of wild biomes and our expanding human ones could provide insight into their roles in maintaining a healthy global biosphere. It's scary that no one knows the limits. Is there a point where further destruction of wilderness ecosystems imperils basic cycling and regulation of the biosphere?

Perhaps more relevant: What kind of Earth do we want to live with and leave for future generations? Do you prefer a world totally dominated by humans and our activities or one with all the other types of biomes, charismatic animals, and iconic plants that the biosphere has birthed?

The inclusion of a wilderness zone could demonstrate the importance of these non-human-dominated systems. In our global biosphere, wilderness areas, unless they have protected status, are subject to encroachment, degradation, resource extraction, and above all, outright conversion for human uses.

There is justified concern about the fate of key biomes like rainforest being converted for agriculture. But forests historically have been lost around the world. The "Great American Forest," from the Atlantic Coast through the South and Midwest, covered most of the United States before European settlement. About 40 percent of the original forest was converted for agriculture and urban use. Of the 60 percent that remains forest, more than 90 percent is used for timber operations. In Europe, half of the western Atlantic mixed forest was cleared for agriculture and 75 percent of the Central European forest. U.S. and European forests are mostly secondary rather than primary forest having regrown after timbering and clear-cutting. In Australia, historic forest loss was 40 percent.[1]

The Assault on the Wild

Human activities now dominate 43 percent of Earth's land area; 40 percent for farming and livestock production and 3 percent for urban use. In 1700, agriculture only occupied 7 percent of the land. Though 95 percent of the world's population is concentrated on 10 percent of the land, just 10 percent of the world is now more than forty-eight hours travel away from human activities.[2] Livestock production, at 30 percent of the planet's land, requires three times as much space as land raising directly eaten crops. Beef production disproportionately occupies far more land than chicken and pig production, though beef supplies only 2 percent of the world's diet, less than dairy, pork, or chicken.[3]

Unprotected wild areas are threatened by the food needs of an increasing population with more demand for livestock products. These trends mean more natural ecosystem land will be converted to agriculture regardless of long-term suitability. Only 20 percent of humid tropical soils are suitable for agriculture. Most rainforest soils are nutrient-poor, infertile, and weathered. The paradox of luxuriant rainforests is that most of the nutrients are contained in the trees' biomass, not in the soil. Nevertheless, rainforest areas are cleared for grazing land and animal feed production.[4]

A New Role for Humans

In Biosphere 2, we wanted to limit human intervention and alteration of our wilderness areas. The small sizes of our biomes meant that food chains would have to be simplified, often excluding the "keystone predator" at the top of food chains. The biospherians would try to consciously fill that duty. A keystone species is a plant or animal with a crucial role in shaping how an ecosystem operates. Without this keystone organism, the ecosystem would be quite different.[5]

Because we lacked herds of savanna-grazing animals in our savanna, we would go in to cut grasses, mimicking what herds do. Grazing or pruning prepares a new season's growth. We intervened in the wilderness when necessary to prevent a serious loss of biodiversity. Consciously acting like keystone predators is a new role for humans. Rather than agents of destruction, we would act for the overall biosphere's health, including managing our atmosphere.

The mixing of the global atmosphere and the reach of basic ecological cycles, like the water and nutrient cycles, as well as global climate change, mean human activities and pollution impact the entire biosphere. Human imagination is still fired by the beauty, diversity, and majesty of nature expressed in the many biome types. A new kind of symbiosis may be emerging, breaking down the strict boundaries of human and "wild."

What was once unimaginable, that humans could have the power to change or manage nature, increasingly becomes a necessity. The U.S. Army Corps of Engineers, by building extensive dams and canals and manipulating the Everglades' water flows, is now a dominant force shaping the huge wetlands. With mounting ecological threats, their mandate has shifted from flood control and resource management to restoring the health of this wilderness area.[6] The Everglades were once called the "river of grass," but now human decisions and devices control river and water flow. Protecting areas of unique diversity and natural beauty has spawned national parks, biosphere reserves, and marine protected areas around the world. Endangered animals bred in captivity and released into the wild exemplify the new symbiosis between humans and the wild.

The Sixth Extinction

Despite these efforts, our biosphere faces grave threats to biodiversity. This human-caused "sixth extinction" is unlike the five mass extinctions in geological history, which resulted from catastrophic "natural" events, like meteor impacts, volcanic eruptions, and natural climate changes. Current loss of species is one thousand to ten thousand times higher than normal rates. Unless trends are reversed, ecologists warn that 50 percent of animal and plant biodiversity could be lost in the next two hundred to three hundred years.[7]

Habitat loss and fragmentation drives these extinctions. Extinctions are happening everywhere, in every type of environment. But rainforests, with their huge biodiversity, face the greatest losses. Extinction threatens half of primates, 20 percent of all mammals, 12 percent of birds, 21 percent of fish, one-third of sharks and stingrays, and one-third of amphibians, including frogs. It's unknown how many insects, by far the greatest number of animals, may be lost because their species numbers have barely been inventoried.[8]

Restoration Ecology and the Challenges of Creating Biomes

Biosphere 2 offered an opportunity to advance restoration ecology, the repairing of damaged natural habitats.[9] Restoration ecology is done from grassroots to governmental levels and is a new scientific discipline. Protection of wetlands and migratory birds has spawned a new application of constructed wetlands to restore degraded natural wetlands. In Florida, constructed wetlands create new wetland areas; when businesses damage or convert existing wetlands, they're required to recreate an equivalent acreage.[10]

The Biosphere 2 project confronted huge unknowns to make its wilderness areas, like how to obtain or create appropriate soils and engineer the necessary environmental backup and support technologies. Even more uncertain was predicting in a tightly sealed man-made biosphere how many species would be needed to compensate for species that didn't survive or were out-competed.

Therefore, in all the land biomes—from rainforest to desert—ecological designers employed the strategy of "species-packing." They selected far more plants and animals than were expected to survive as the ecosystems matured. They endeavored to include several that might play the same role in food chains or occupy similar microhabitats. Wind or insect pollination introduced another complexity to choosing which plants to include.

For the ocean and marsh, species selections were made, and whole chunks of living communities were introduced. The mangroves and other marsh vegetation were transported to Biosphere 2 in wooden boxes. This ensured Everglades soil and microbiota also made the journey. Corals from the Yucatan, Mexico, were collected and transported in lugs on specially designed tractor trailers. The lugs included seawater teeming with phytoplankton and other marine life.[11]

The decision to include soil in the agriculture and wilderness biomes was driven by appreciation of their immense life power. Underneath one square foot of fertile soil live about a million roundworms, five thousand earthworms, and five thousand insects and mites. Microbial soil life is even more intense. A thimble of soil contains thirty thousand protozoa, fifty thousand algae, half a million fungi, and billions of bacteria—the vast majority of which have never been seen nor identified.[12] Unlike more complex organisms, bacteria can exchange genetic material across species boundaries, independent of reproduction. This helps microbes quickly adapt to changing environmental conditions. For example, if a bacteria manages to survive, or if it uses a previously toxic chemical, it can pass those genes to unrelated bacteria.[13]

Microbial life permeates our global biosphere and makes it work. A variety of microhabitats in each biome was aimed at maximizing soil diversity to ensure a full range of microbial function. There were more than thirty soil types throughout Biosphere 2. Dr. Robert Scarborough, an expert in Arizona geology, was a major consultant on building these soils.

Microbes in water environments play equally important roles, so a wide range of aquatic environments were included. Aquatic microbial life is as unstudied as soil; science lacks even a rough understanding of all the relationships and functions of water microbes. What we do know, as Dr. E. O. Wilson, the Harvard biologist, wrote in "The Little Things That Run the World," is that we need them, but they don't need us.[14] Biosphere 2 started its first closure with about 3,800 species of plants and animals. But our world also included mind-boggling and incalculable numbers of microbial organisms and other soil and aquatic life. We hoped they would help run our minibiosphere like they do our global home.

Animal selection was fraught with risk and the unknown. Scientists rarely know the complete diets of wild animals, especially herbivores. Even more difficult to predict is what particular animals, including insects, would feed on in synthetic ecosystems. Behavior patterns also determined whether a candidate species was selected.

We wanted to have hummingbirds. But Dr. Peter Warshall, our savanna designer in charge of vertebrate selection, soon found that many have a mating flight, with some ascending more than one hundred feet, a height that could not be accommodated inside Biosphere 2.

Imagine the fun calculating how many flowers nectar-feeders need and what species would be flowering in differing seasons in Biosphere 2, or determining how many insects an insectivore needs to eat each day. Lack of UV light excluded daytime lizards and honeybees, which need the light for successful navigation. A thousand similar questions occupied our wilderness biome ecologists for several years as species lists were drawn up.

Logistics and Quarantine Regulations

Once wilderness species were selected, we faced logistical challenges: how to get the plants and animals, obtain permission to import them to Arizona, and house them in support greenhouses until it was time to plant or introduce them to their part of the wilderness. The Arizona state government was extremely cooperative. They designated specially designed greenhouses we built on-site as quarantine areas to inspect and hold new acquisitions to exclude unwanted species or pests.

Quarantine regulations are a typical paradox. Though intended to protect nature, the chemicals specified in the regulations to control pests and diseases are extremely toxic. We were advised to take the easy way out and get rid of the nasty stuff in a toxic waste dump, but that was not an ecotechnic approach or one compatible with Biosphere 2 goals. Instead the project invested in research, later patented, to design a facility to de-toxify the wastewater containing residues from the mandated poisons.[15]

Food Chains: Galagos but Not Tigers and Elephants

Given space constraints and food chain realities, the largest animal in our wilderness biomes was the galago, or "bush baby," a small monkey-like, tree-dwelling animal from Africa. Weighing just two and a half pounds at maturity, they could survive on fruit and insects in Biosphere 2. They had done well and multiplied in our research greenhouses before closure. There was supplemental "monkey chow" for them just in case.

Since we wouldn't have pets, these lively prosimians would be our primate cousins, offering entertainment and a kind of companionship. They were known to be friendly toward humans without losing their wildness. Night-active, we expected them to traverse the tree lines from rainforest through

Figure 35. A Biosphere 2 galago (bush baby). They are African tree-dwelling, night-active prosimians (related to monkeys), and the largest of the wild animals inside Biosphere 2.

savanna to thorn scrub. They had a true monkey-like curiosity and fully explored our little world more widely than expected. They traversed our technosphere basements, and one even reached our orchard, taking advantage of a door left ajar.

After Biosphere 2 captured the world's imagination, the educational value of our wilderness biomes was reflected in innumerable class projects in which students designed a biosphere, which provided stimulating ecology lessons. We explained why you couldn't have an elephant, tiger, or lion in a half-acre rainforest. A minibiosphere is not a zoo or aquarium where an unlimited amount of meat for predators, bamboo for pandas, grass for zebras, and fish for sharks can be supplied. A food chain has to be built to provide needed food on a continuing basis. Discoveries like this were great teaching moments for so many students around the world and great fun for

Figure 36. Linda meets visitors to Biosphere 2 via two-way radio (photo by C. Alan Morgan).

us biospherians to be able to interact with students from grade school to college level.

Defending Wilderness from Agri-Imperialists

Despite our hunger, there was never any serious attempt to convert our wilderness to food production. It helped that Gaie was fiercely protective of the ocean and marsh and Linda equally so about the terrestrial wilderness. We did insert a few more edible plants, especially in changes after the first closure, but Linda was vigilant in resisting any incursions by "agri-imperalists."

She told Kevin Kelly, editor of the *Whole Earth Review (CoEvolution Quarterly)*: "The wilderness is designed to provide an *occasional* bowl of bananas, passion fruits, papayas, and various tropical fruits seasonally. Wilderness will continue to be wilderness as long as I'm in."

I concurred: "If we say, 'Let's plant a few tubers of taro in the rainforest,' Linda is happy to allow that. But if we go in there prospecting with a spade, she's out front beating back the competitors."[16]

We also knew the wilderness was vital in keeping our biosphere healthy. Unlike the farm where crops were harvested and soil disturbed, the wilderness plants kept growing, assisting us in atmospheric management. I told Kelly for his article at the one-year point of our closure: "If not for the wilderness areas, we wouldn't have made it through Christmas."

Kelly also noted, "Wilderness offers a deep keel to steady the great cycles."

We fell in love with the magic of our baby biomes—and knew what an effort had gone into creating them.[17]

9
Building a Rainforest

The Rainforest Team and Plan

THE RAINFOREST DESIGNERS were tasked with doing the near impossible: make a rainforest from scratch and ensure it flourishes in a southern Arizona greenhouse environment with harsh desert sunlight coming in from all the glass sides, and make half an acre that captures the feel and ecological function of the biome with the world's greatest biodiversity.

The task was headed by Dr. Ghillean Prance, vice president of the New York Botanical Garden (NYBG) and a world expert in the plants of the Amazon and the new world tropics. Dr. Mike Balick, director of the Institute for Economic Botany at NYBG, also helped with rainforest design. Later, when Prance was selected to be director of the Royal Botanical Gardens in Kew, England, we wound up with help from the staffs at two of the world's greatest botanical gardens.

Wilderness areas are replete with natural variation, offering mosaics of differing landscapes. Each has distinct soils, waters, and particular groupings of plants and animals. The task in Biosphere 2 was to compress that spatial variety into a bonsai size. During the design workshops, a plan was devised to make that half acre as diverse and full of differing rainforest ecotypes as possible.

Ecotones—the boundaries where one type of community changes into another—would be critical. A mountain of synthetic rock modeled on a tabletop formation, a "tepui," found in Venezuela in the Guiana Highlands of South America would dominate its north side.[1] It would include planting pockets up its sides, mountain terrace forest areas on its edges, and a cloud forest with misters at the top to try to replicate continual fog.

The Larson Company of Tucson, pioneers in replicating the look of natural rock faces with special forms and painted concrete, built the rainforest mountain. They constructed a thin steel frame sculpted like natural rocks covered with a thin layer of concrete to minimize weight. They also created other large structures in Biosphere 2 like the thirty-five-foot savanna cliff and desert caves.

The rainforest, when planted, was a spectacular place to work and visit for fun. A pool at the top of the mountain overflowed twenty-five feet to a waterfall dropping into a splash pool and then a larger water basin (Tiger Pond) at the mountain base. The water then flowed through a meandering várzea, a seasonally flooded area with a riparian plant community, where

Figure 37. Building wilderness biomes about two years before closure in 1989. Tallest structure is the rainforest mountain, the savanna cliff is on the left, and the beach already has two coconut palms.

distinctive trees like Phytolacca would grow.[2] The lowland rainforest fanned out at the base of the mountain, filling a valley to the south. The stepped pyramid space frame towered overhead some seventy-five feet above the soil we walked on.

To protect the rainforest interior from harsh sunlight coming in from the sides, another ecotone was designed. This "gingerbelt" was filled with plants from the Zingiberacae family, including ginger, cardamom, and turmeric, heliconias (the beautiful bird-of-paradise flowers), bananas, and plantains. These types of rainforest plants can withstand intense sunlight, unlike most rainforest species.

To protect the lowland rainforest from the harsh Arizona sun, an ecological succession was planned. Fast-growing trees like Cecropia (with huge silvery leaves), Ceiba, and Leuceana would form the first canopy. Later, the slower-growing, shade-tolerant, mature rainforest trees would supplant them. The rainforest would provide a laboratory for studying methods for protecting our natural rainforests that face harsh sunlight when bordering areas are cleared.

The bamboo belt formed another protective ecotone on the southern edge of the Biosphere 2 rainforest. Fast-growing bamboos provided another landscape type as well as preventing salt aerosols from the adjoining ocean from reaching the rest of the rainforest.

Collections for the rainforest included species from Brazil, Venezuela, Belize, and Puerto Rico. Linda went to Guyana to collect cloud forest species. She contracted malaria in the line of duty. The Missouri Botanical Gardens donated some large specimens, as they were redoing their Climatron rainforest exhibit in St. Louis. Other seeds and plants came from botanical gardens, nurseries, private collections, and the ecotechnic project in Puerto Rico's mountain rainforest. Each of the rainforest habitat types had a distinctive array of plants, which included vines, ground-cover plants, epiphytes (air plants), shrubs, and trees to fill several layers of canopy.

Animals included the galagos, a few snakes, lizards, frogs, and tortoises. There were intentionally introduced insects, including small, dark cockroaches to build food webs and help with recycling dead organic matter. Mosquitos, which don't feed on people, were discussed but later dropped from the selected species list.

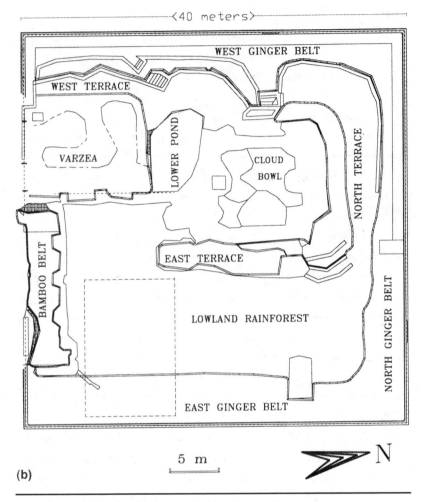

Figure 38. Schematic of Biosphere 2 rainforest ecozones.

Mapping and Measuring

The species-packing strategy resulted in more than 1,800 plants from more than 300 species forming the initial planting in the half-acre Biosphere 2 rainforest. Most were relatively small plants, though a few trees were eight to ten feet tall. A meticulous mapping and measurement of each plant in the rainforest (and every biome in Biosphere 2) was done. Dr. Kristina Vogt and Dr. Dan Vogt of the School of Forestry and Environmental Studies at Yale University headed up the research program. In total, more than eleven thousand plants throughout the facility were measured so we could calculate our initial biomass. Each plant was also mapped so we could track changes over time.

Some, like H. T. and Eugene Odum, saw Biosphere 2 as the greatest experiment ever carried out to study ecological self-organization.[3] In addition, Kristina and Dan worked out the methodology for the program to archive soil samples from all of our biomes before and after the two-year closure so future research can correlate ecosystem changes with soil changes. Dr. Tom Siccama from Yale and Dr. Ernesto Franco from Michigan State University worked on soil analyses as well, documenting changes that occurred during the two years in the rainforest and farm.

The Biosphere 2 rainforest flourished. Trees and other plants grew rapidly. The rainforest increased its aboveground biomass by 400 percent during the two years. These estimates were based on measurements done by Linda and myself during the two years.[4] After the two-year experiment, teams remeasured and mapped every plant in all the Biosphere 2 biomes during the period before a second closure experiment was started. This invaluable data set has never been analyzed, and it is unknown exactly who controls them.

Perhaps the trees grew too rapidly because of unusual conditions in Biosphere 2. Light was low since half of the sunlight was blocked by the space frames and internal shading. Summer days were longer than rainforest trees normally experience in the tropics where seasons do not differ as much. Growth rates may have also been increased by higher CO_2 levels. As a result, plants throughout Biosphere 2's wilderness areas tended to be what ecologists call "etiolated." That is, they grew tall and thin compared to plants with more sunlight. This particularly affected trees, which also lacked "stress wood," higher strength tissue produced with exposure to high turbulence winds. This led to some drooping of trees, partly compensated for by tying treetops to the overhead space frame ceilings.

A method of circulating hot air from the rainforest mountain and overhead space frames was nixed because of budget concerns. The high temperatures limited tree growth above fifty feet and made it impossible to maintain true cloud forest plants on the rainforest mountaintop.

Plate 1. Biosphere 2 facility in southern Arizona. The white domed structure is one of two "lungs" (expansion chambers) connected to the main structures by air tunnels. In the distance is the Biospheric Research and Development complex, where much initial system research, biospherian training, and raising of plants, animals, and insects occurred. Biosphere 2 population: 30,000 tons of soil, 3,800 species of plants and animals, 8 humans (photo by Gill C. Kenny).

Plate 2. Biospherians for first closure experiment, 1991–1993. Back row from left: Linda, Taber, Sally, Laser. Front row from left: author, Jane, Gaie, Roy.

Plate 3. Laser, manager of our technosphere, in the technical basement of Biosphere 2 (photo by Peter Menzel).

Plate 4. Linda planting in the cloud forest. Below, the lowland rainforest grew prodigiously.

Plate 5. Lower savanna from upper savanna ridge. Thorn scrub in distance.

Plate 6. Mud, mud, glorious mud! Farm crew planting rice seedlings. Taber is in foreground, then Jane, and near windows in back, Gaie and Laser.

Plate 7. A portion of the Biosphere 2 main farm area showing some of the sixteen cropping areas. We also grew small Leuceana trees for high-protein fodder. Planting beds also cover two air vents (photo by Abigail Alling).

Plate 8. Happy crew in front of a heavily laden feast table including roast pig (photo by Abigail Alling).

Plate 9. The care and weeding of the coral reef ocean. Gaie checking the reef and removing algae. Eleven pounds a week were removed, helping the corals retain access to the sunlight they need for photosynthesis.

Plate 10. Party in the command room, our shared central office space (photo by Roy Walford). Note the large birthday cake topped with bananas that Taber is sampling.

Plate 11. Sharing some of the stories with Oleg Gazenko and John Allen. Gazenko was the unofficial confidante and psychologist for generations of Russian cosmonauts. His opinion was that he'd seen much worse conflict amongst Russian space crews! His observations indicated the biospherians had fully adapted to our environment (photo by Marie Harding).

Plate 12. Linda and I remeasured plants to track rates of growth and increase in biomass in all the wilderness biomes (photo by Abigail Alling).

Plate 13. With H. T. Odum in front of the first Wastewater Gardens in Akumal, Mexico, in 1996 (photo by Gonzalo C. Arcila).

Plate 14. Wastewater Garden at Las Casas de la Selva, Puerto Rico. Sally, who then managed the ecotechnic sustainable tropical forestry project, inspects some flowering plants.

Plate 15. The marsh at the end of the two years. Growth was phenomenal.

Plate 16. First steps, first breaths in a new world. The biospherians emerge led by Gaie on September 26, 1993. Bill Dempster, Biosphere 2's director of systems engineering, stands by airlock (photo by D. P. Snyder).

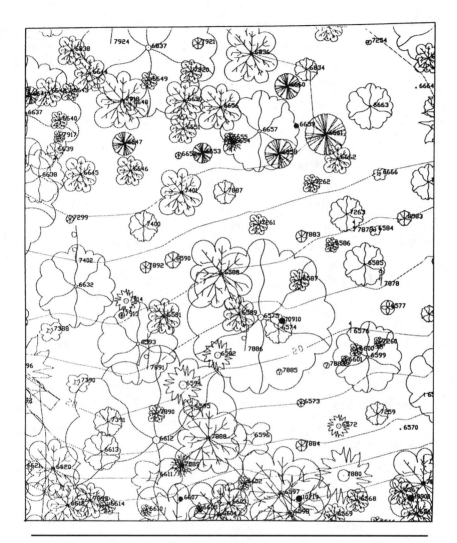

Figure 39. A section of the mapping and measurement of the Biosphere 2 rainforest in 1991.

The biosphere's rainforest certainly achieved the goal of replicating the feel of a natural rainforest. The high humidity, dense vegetation, and shade canopy that quickly formed gave the feeling of being in a whole other realm. The selection of plants, apart from habitat differentiation and building food chains, were also designed in part to give special treats for the humans. The young coffee trees and some yerba mate plants (the caffeine plant of choice for many South Americans) were to provide stimulating drinks. Bananas, plantains, and the ginger family herbs helped spice up some of our meals. The rainforest also delighted us, our world's first inhabitants, with gorgeous flowers and the aromas of the wet tropics.

Its dense "jungle" cover allowed for privacy and special spots for crew members to get away from the madding crowd, even if that crowd numbered just seven other people. Sometimes it was like one of those zoo exhibits where visitors try to see the well-camouflaged animals in the glass exhibit. But in this case, the biospherians were the elusive animals. We could tell that visitors along the glass had no idea that we were enjoying our public privacy a stone's throw from where they stood.

Defending Biodiversity

We biospherians interacted with the rainforest as atmospheric stewards, cutting ginger belt plants capable of fast regrowth to help control CO_2 on Team Biodiversity. We also worked as keystone predators. We especially kept a careful eye on our intermediate and top canopy trees that would supplant most of the early successional trees. We pruned back plants that excessively shaded them and vines if they were too numerous and hitchhiked a ride on the trees to the sunlight overhead.

But it was the morning glories that proved the most daunting foe and took the most human intervention time. They were planned for some of the soil pockets on the rainforest mountain. We were glad to see them cover the mountain with greenery, but the morning glories didn't stop there. They invaded mountaintop and terrace areas. Since water proved no barrier, they crisscrossed our ponds and pools, climbing out on the other side to conquer the várzea. They also tried to cover everything in the lowland forest. They were even poised to invade the savanna from the edge of the várzea when we mobilized against them! Crew after crew worked on pruning them back. Morning glories put down roots as they grow laterally, making it hard to figure out where the beginning or end of them is. Pulling up morning glory vine felt at times like rolling up endless spools of electric wire, for their stems are quite slender.

At the end of the two years, during the transition before a second crew began a closure experiment, we removed the Leuceana trees from the rainforest. Some had grown from four- to five-foot tall saplings to robust trees as high as thirty-five or forty feet! Their removal would allow the mature rainforest trees to get some extra light, now that they were strong enough to occupy the top canopy. The bamboos had also grown from four feet to more than thirty feet tall. The young Phytolacca trees in the várzea floodplain area grew so large you couldn't put your arms around their trunks.

After Biosphere 2, Linda received her doctorate in environmental engineering with H. T. Odum at the University of Florida, as I was doing. Since her dissertation concerned the factors promoting rainforest biodiversity, her work has provided us with great information on how the Biosphere 2 biome developed.

As expected, rainforest diversity and the number of plants decreased, but hardly to the catastrophic levels some had feared. H. T. Odum thought we'd lose about 80 percent of our biodiversity by the end of the first two years. Linda ventured a 20 percent loss. Neither was right in the rainforest. In the inventory after the two-year closure, the 288 original species declined by about 40 percent to 181, and 45 percent of the original plants survived for two years. Despite loss of almost all pollinating insects to voracious ants, many of the rainforest plants flowered and seeded.[5] The loss of almost all our pollinators was attributed to a type of generalist ant, *Paratrechina longicornis*, called the "crazy ant" because of its erratic movements. This stowaway is frequently found in greenhouses. It became prevalent through the terrestrial biomes. Only insects that were safe from predation by crazy ants survived the two-year experiment.[6]

Galago Dramas

As expected, the galagos provided lots of entertainment and bonding. We took night watches on a rotation, and galagos were frequently encountered or heard calling during our rounds. The calls of the galagos resounded through the wilderness, cheering up night-watch patrols and biospherians out for an evening in the wilderness biomes. More than a few times, galagos would perch in a savanna tree overlooking us at dinner. They were curious, and so were we.

The two adult females had fought when they were being raised in the project greenhouses prior to closure. While our primate consultants thought Biosphere 2 offered enough habitat to allow them both a territory, this was not the case. The alpha female inflicted some nasty cuts on her less dominant competitor. We were finally obliged to build a spacious cage in the orchard so the two could be separated for the last year of our time inside.

There was a tragic death as well. The galagos, driven no doubt by our shared primate sense of curiosity, made the technosphere basement a place of exploration as well as the green biomes. A younger galago was electro-cuted where he found access to an unguarded ceiling junction box. After-ward, Linda and Laser carefully surveyed the entire technosphere, galago-proofing any places of possible harm. We also had the thrill of a baby galago being born to our alpha male and female during the two-year closure.

Special moments were shared by several of the crew, including Gaie and Linda. While seated in the rainforest or beach, a galago would silently come up and sit next to them. Linda felt like it was a visit from a "little sister."[7] They would sit together for a while, and then the galago would go on its way.

At the end of two years, we introduced another smaller group of rainfor-est plant species, including a few with fruit, to enliven dinner plates. This was part of our strategy that allows for sufficient time and ecological self-organization before a created biome could develop to its maximum diver-sity. Size limitations would undoubtedly prevent Biosphere 2 from having anything but a tiny fraction of the extraordinary biodiversity and mature canopy that characterize healthy, natural primary rainforests. It was, after all, just half an acre in size. But it remains an open question as to how much biodiversity can eventually be sustained. By theory, as the system matures and stabilizes, less human intervention may be required for invasive species.

Linda concluded, "Initial [rainforest] dynamics seem to approach those in other tropical rainforests. The different light and CO_2 regimes may alter biogeochemical cycling; hence the Biosphere 2 rainforest is a suitable plat-form for innovative research."[8]

The rainforest of Biosphere 2 remains a potent tool for studying our most biodiverse and threatened biome. Because environmental vectors can be controlled so well, the Biosphere 2 rainforest is still being used as we had

Figure 40. Sir Ghillean Prance and his wife, Anne, inspect the rainforest during transition. In rear: John Allen. Dr. Yong Dan Wei stands next to me. Prance's strategies for planned succession and protecting the inner rainforest worked extremely well.

hoped. It serves as a comparison for studying Amazonian rainforests, and is also used as an experimental tool for studying response to drought, elevated CO_2, and the increased warmth climate change may bring.

The Planet's Rainforests

The difficulties we faced paralleled troubling dynamics in our world's rainforests: fragmentation caused by rainforest clearing. An early study looked at the impact on biodiversity and ecosystem health when rainforest patches were 2.5, 25, and 250 acres in size. As might be expected, the larger areas proved most resilient. The smaller areas had many more "edge" effects, less habitat variation and drying winds penetrated deeper. As a result, many collapsed, unable to sustain themselves. Similar studies around the world and in different biome types have confirmed these results.[9]

Saving rainforest areas such as the rapidly shrinking Amazon and areas in southeast Asia, Indonesia, and Africa should not just rely on protected areas. Creation of national parks is needed, especially to ensure the preservation of especially biodiverse and unique rainforest habitat regions. But all too often, simply declaring an area protected does not make it so. Corruption and greed lead to widespread illegal logging by the wealthy and influential. Excluding desperately poor people leads to encroachment. Often the most effective way of protecting rainforest is to couple it with a plan to improve the local economy by providing jobs to native and local people in their protection and with sensible resource utilization.

A great example of this is in Nepal, where former poachers among neighboring populations were employed as park rangers to protect the Chitwan National Park area. The land was one of the largest remaining rainforest areas of the distinctive Terai mountain foothills of India and Nepal. It had been largely cleared of mosquitos with the pesticide DDT (dichlorodiphenyltrichloroethane) to combat malaria before the harmful environmental impacts of this chemical were discovered. Then the government decided to protect the land in the early 1970s because it supports important populations of Bengal tigers and Asian one-horned rhinos. Villagers who were causing damage by overcutting forage grasses are allowed in only at certain times of the year, so they are not simply excluded. The restored and managed areas now supply the villagers with more forage than previously since the ecology is now more robust, among other socioeconomic benefits.[10]

Amazonian and other rainforests look natural to outsiders who don't understand that they have been managed and utilized by their indigenous populations for millennia. A study by the Institute of Economic Botany found that more than 80 percent of the plant species in the Amazonian rainforest adjoining Indian villages are harvested for food, medicines, fiber, oils, insecticides, and other uses.[11] Most "natural" rainforest is really a synthetic ecology because it was shaped and has co-evolved with the human cultures that depend on it.[12]

One way that traditional peoples impacted tropical forests (and savannas with trees) is through slash-and-burn agriculture. This approach can allow for farming despite poor soils, dense vegetation, and low levels of soil nutrients. This system does not damage the land as long as there are adequate resting periods between use periods. But if population pressures increase and the interval between uses is shortened, as has been happening, the consequences can be devastating. It is estimated there are still some three hundred million people practicing slash-and-burn agriculture.

Researchers are working on ways to adapt slash-and-burn, small-scale rainforest farming to be more compatible with long-term health of the forests. One approach uses strips of native leguminous rainforest trees to protect crops and add nitrogen, along with mulch cover to protect the soil from erosion. This provides traditional farmers in the rainforest with fuel wood and consistent, sustainable crops without soil damage while preserving, not destroying, rainforests for short-term gain.[13]

Short-Term vs. Long-Term Economics

That rainforests must be cut down for economic reasons has long been disputed. Landowners in the Amazon only get $60 per acre per year for converting rainforest for grazing and $400 per acre for the one-time extraction of timber. By contrast, valuable foods, medicines, rubber, and other non-timber uses yield $2,400 per acre, which can be sustained long-term.[14] Utilizing the rainforest in this way conserves the rainforest's ecological diversity and other benefits to our global biosphere. Up to 50 percent of the Amazon's rain is generated by the standing forest and is lost when large areas are cleared.[15]

Almost 40 percent of the world's medicines come from plants; 25 percent from the rainforest. Ongoing forest loss will mean many potential medicines and other valuable plants will be forever lost before they are even found. Only 1 percent of Amazonian plants have been screened for medicinal activity. Half of global rainforest area is already gone. Twenty percent of the vast Amazon has been cleared in the past fifty years. We continue to lose two hundred thousand acres of rainforest *each day*.

Not only is the rainforest being lost, but so are the traditional tribes. The native peoples of the Amazon are its greatest ethnobotanists. One village used two hundred different plants in their surrounding rainforest as medicinal substances. In the Amazon as a whole, it is estimated between 1,500 and

2,000 fruits are used by Indian tribes. Contrast that with just two hundred fruits in the diet of the rest of the world. With the loss of those people, we are losing the key to their living drugstores and food baskets. A tribe in the Colombian Amazon can identify all their tree species even without flowers and fruits, something no university-trained botanist can do.[16] The indigenous population of the Amazon has been decimated from almost ten million about five hundred years ago to less than 250,000 currently.[17]

One of the demonstration projects of the Institute of Ecotechnics is in the wet tropical forest of Puerto Rico. Las Casas de la Selva has been working, with the support of the government, since the early 1980s to demonstrate that secondary forests can be managed to maintain forest cover while line-planting, which leaves the forest intact between planting strips, valuable hardwoods that can provide long-term economic benefits. Secondary forests are increasing everywhere because of the previous clearing and abandonment of farm and grazing land. Puerto Rico itself has the fastest rate of reforestation in the world. Sustainable timber can be harvested from secondary forests by enriching them and using selective "liberation thinning" to accelerate the otherwise slow growth of valuable native timber trees. This could alleviate some of the pressure on primary rainforests.[18]

Fair trade products like coffee, Brazil nuts, and others that are certified deforestation-free are gaining traction as a way to support indigenous people and their use of the rainforest. Pharmaceutical companies face pressure to stop "bio-piracy" and to share the profits of their medicines with the native peoples who often guide medicine hunters to the plants they know are effective.

Businesses and governments face pressure to reduce and stop deforestation. Global climate change activists support plans to preserve rainforest as a carbon sink rather than the carbon release associated with clearing and conversion to agriculture. A tax on carbon will increase incentives to preserve rainforest. Improving the resource efficiency and productivity of lands already converted; empowering local stakeholders, including indigenous peoples, in their rights to manage their rainforest; restoration of damaged, degraded ecosystems; and exerting consumer pressure to avoid products derived from rainforest destruction and support sustainable non-timber products can all help preserve and restore the rainforest.[19]

Community and Small-Scale Sustainable Forestry

This is already happening to a surprising extent. In developing countries, 27 percent of forests are community owned or managed. When ownership is in local hands, management tends to be more sustainable, since long-term use and secondary uses are more important than short-term clear-cutting. Community based forestry management results in far more and continuing employment, generating 90 percent of total forest income and 50 to 90 percent of jobs.

Better recognition of community-based forestry enterprises and return of land to indigenous and local ownership would reverse the current trend toward allocating timber concessions and assigning protected status without regard to historical tenure and land rights. In effect, "they were a land-grab by the state under the guise of national (or international) interest."[20]

The rainforest is not being destroyed for rational economic reasons. It is being destroyed for the short-term profit maximization of the rich and powerful, the elite of their own countries, and the international market. This is often combined with an attitude of contempt for traditional indigenous people.

As the battle for the future of our rainforests clearly shows, issues of social justice and the empowerment of people, including indigenous populations, are essential for a more viable future for them and our biosphere.

10

What's a World Without an Ocean?

High-Elevation Desert Arizona

CREATING A LIVING OCEAN with a tropical coral reef was undoubtedly the most challenging project of the wilderness biomes. Keeping it healthy required hard work, ingenuity, and passion. Tropical coral reef exhibits at aquariums are always difficult to maintain. They require periodic exchange to bring in fresh ocean water to flush out the buildup of nutrients and replacement of dead corals. Biosphere 2 needed an ocean for its function as a laboratory for global ecology. Arthur C. Clarke, the science fiction writer who began his career as a marine biologist, noted that since two-thirds of its surface is water, "How inappropriate to call this planet 'Earth,' when it is clearly 'Ocean.'"[1]

Coral reefs have come to symbolize the majesty and sublime beauty of our oceans. They are among the oldest ecosystems in the world. The Australian Great Barrier Reef extends 1,200 miles and has been growing for more than five million years. Coral reefs cover 110,000 square miles and are vital to sustaining fish and marine populations and the livelihoods of millions of people in poor tropical countries. A single reef can harbor 4,000 different species, a biodiversity more than one hundred times greater than the open ocean.[2]

There were many reasons people said building our own miniature ocean couldn't be done. Tropical coral reefs are found in the near-equatorial regions of the planet, not at the Biosphere 2 location, where seasons vary greatly.[3] They are at sea level, not at the 3,900-foot elevation of the Biosphere 2 site. Arizona is a long way from a tropical ocean, and corals are notoriously difficult to transport. They need light and water circulation while being moved. How can you find the right kinds of limestone for the reefs in arid Arizona? To keep a coral reef healthy, there can't be a buildup of nutrients.

Biosphere 2 found great resources in working with Dr. Walter Adey and the Marine Sciences Laboratory of the Smithsonian Museum of Natural History. Adey was a pioneer in creating ecological mesocosms, small versions of natural systems, for public education and research, and collaborated with the project on both the ocean and marsh systems. His group had already created a coral reef mesocosm exhibit for his museum in Washington, DC, and another in Australia showcasing the ecology of the Great Barrier Reef corals. They had developed ways of keeping nutrients from building up by miniaturizing algae tanks to serve as a biological cleanser of the waters.[4]

Earth's geological history helped. Arizona and good portions of the southwest United States may be high and dry now—but there have been numerous shallow oceans covering them during the last two billion years, so ancient marine limestone could be found locally as a foundation for the coral reef ocean.[5]

Biosphere 2's ocean water was a complex synthesis. One hundred thousand gallons (about one-tenth its volume) came from coastal waters near Scripps Oceanographic Institute in La Jolla, California. A first shipment was fortunately rejected by alert eyes noticing that the trucks had previously carried fuel and oil. If that shipment hadn't been rejected, Biosphere 2 would have begun with a polluted ocean, analogous to far too many modern coastal waters near industry and cities. We made sure the next shipment came in clean, off-duty milk trucks! This ocean water brought in myriads of natural ocean microbes and small critters and phytoplankton—our "yogurt" start. The rest of the ocean water came from well water mixed with Instant Ocean, the mix of ocean salts used by aquariums.

The ocean was about 150 feet long, 60 feet wide, 25 feet at its deepest, and 14 feet at the shallow end. It held about one million gallons of water. The gentle waves created by vacuum pumps lapped on our beach shore. In addition to the seashells and sand imported from the Gulf Coast, it was planted with typical salt-tolerant beach plants. A few young coconut palm trees completed its tropical feel.

The Pilgrim Corals

Friends in Mexico helped us secure government cooperation and permits for a careful collection of corals and other marine life from the Yucatan coast near Akumal. Teams of divers from the RV *Heraclitus* rendezvoused with divers who flew down from the biosphere project in Arizona. Collecting was managed by Gaie, our marine biologist. She would manage the ocean and the marsh once they were in their new home while also serving as associate director of Biosphere 2 research.

The corals had their depth noted and their location in fore reef, flat middle reef, or rear reef lagoon so they could be similarly placed in our ocean. The collections included four dozen species ranging from soft, waving corals to brain corals. More than forty types of tropical reef fish, crabs, lobsters, shrimp, and seahorses were also collected. The trucks also carried Caribbean sand and limestone to complement the local Arizona limestone rock.

Each truck transformed from an ordinary large tractor trailer into an aquarium on wheels. Their cargo area had bright lights overhead and pumps and pipes to circulate water between the tanks holding the living coral. They even had a small algal turf scrubber to remove nutrients. We received a Mexican police escort on the 1,800-mile journey from the Yucatan to the world's newest ocean. A five-part Mexican television series by Tiahoga Ruge on Biosphere 2 was aptly titled *The Pilgrim Corals* (*Los Corales Peregrines*).

Once installed, Biosphere 2 had the world's largest artificial reef. There were forty-nine species of hard and soft corals and dozens of tropical fish. Now, we'd have to work to keep them alive and healthy.

The Battle to Keep Our Ocean Healthy

The first issue was ocean pH and its impact on our coral's ability to make new tissue and reproduce. We knew Biosphere 2's atmosphere would have a far higher CO_2 concentration, though it was unknown how high it might go. The absorption of carbon dioxide into ocean waters lowers their pH, making them more acidic. Corals are usually found in waters with a pH around 8.2. It was almost completely unknown then what impact straying from those levels might mean.

Maintaining ocean health was a priority throughout the two years. Weekly water analyses tracked critical parameters. A shut-down of the wave machine for mere hours presented danger, so there was a 24/7 response to vacuum pump alarms by Gaie and Laser until necessary repairs were made.

The ocean team added carbonates and bicarbonates periodically to help raise pH. The marsh was disconnected from the ocean early in November 1991 (six weeks into closure) because marsh water was pH 7.6. Calcium compounds, a vital component for new coral tissue, were also added to ensure adequate availability. With these efforts, pH never fell below 7.65 and mostly was 7.8 or 7.9 during the two years. In small aquariums, this might have proved fatal, but a surprise was that the corals survived and reproduced in Biosphere 2 at lower pH levels than are found in nature.[6]

Keeping the ocean low in nutrients was also critical. At first, we used the algal turf scrubber system. In a room inside the savanna cliff-face, high-intensity lights shined on shelves of stacked plastic trays with a tipping bucket that sent water pumped in over an area with natural marine algae. The system emulates the water cleaning that algae do in coastal wetlands. Every week, two biospherians cleaned some of the trays, scraping off the algae so that it could regrow. The algae removed nutrients from both ocean and marsh water. It was dried in an oven to reduce volume. In the future, when ocean and marsh might need more nutrients, the algae could be reintroduced.

Scraping algae was the lousiest job we had to do, and it required ten to twelve hours per week. It was an elegant design to use a natural filtering system with diverse and some beautifully colored algae. But the algae scrubber room was loud, salty, and sticky, with harsh lighting as well. We were relieved when our marine consultants recommended "protein skimmers" as their replacements.[7] These could remove nutrients more effectively with far less labor. The skimmers were long PVC pipes with aerators that bubbled organic materials containing nutrients into foam at the top, which could be removed.

Figure 41. The lousiest job, but we did it to safeguard the health of the ocean. Taber cleaning mats in the algae scrubber room.

These protein skimmers were constructed inside with spare materials in our workshop—another case of human ingenuity and response to safeguard our wilderness biomes.[8]

The ocean team in Biosphere 2 became keystone predators when three types of introduced lobsters threatened the snail population. Snails feed on and help keep algae in check. Two species of lobster were removed to improve food chain balance. The decision whether to eat the removed lobsters divided the crew a bit acrimoniously. A few refused to eat them for fear that heavy metals might have leached into the ocean water. The others thought this concern was exaggerated and chowed down on our two lobster meals.

Parrotfish (beautifully colored fish that take occasional bites out of corals) and squirrelfish (which feed on the young of other fish) were also reduced in number to balance food chains. Bright-red fire worms, which feed on corals and anemones, were stowaways, probably tiny, when they hitchhiked a ride on our coral trucks. Divers removed them by the hundreds, using gloves since the fire worms deserve their name, until they were no longer a problem.[9]

Coral reefs—even new, synthetic ones—also require weeding, especially when nutrients were higher than desired and algae thrived on our reefs. Algal growth reduces the light corals need. The corals were already adjusting to lower levels of sunlight than they receive in tropical oceans. Gaie, Laser, and occasionally other divers went in to weed algae off the reef, ocean bottom, and other surfaces. Gaie also moved some corals higher on the reef during winter months and put them lower in the longer-daylight months.

To monitor the health of the Biosphere 2 corals, some innovative approaches were employed that allowed the help of our network of outside consultants. Dr. Phil Dustan of the College of Charleston was a pioneer of video monitoring using special equipment to assess photosynthetic activity. That coupled with human repeat observation—the biospherians as naturalist observers— allowed us to track changes in the health of our corals. Dr. Robert Howarth and Dr. Roxanne Marino of Cornell University consulted on the challenges of ocean chemistry.

The Scorecard after Two Years

Dr. Judy Lang, then of the Texas Memorial Museum, and Dustan helped map the coral reef at the end of the two years. More than 1,220 hard and soft living corals were found, only one species of coral was lost, and there was reproduction with 87 baby coral colonies identified.[10] Gaie recalled, "When we closed, there wasn't a single scientist we were working with who thought the coral reef would survive."[11]

Nevertheless, it had been a struggle—and the combination of lower light, lower pH, and water quality concerns may have been the cause of about a one-third reduction in tissue size during the two years. A brief outbreak of white band disease, which causes corals to lose their vitality and ability to photosynthesize, occurred but did not widely spread. Dr. Don Spoon of Georgetown University discovered a marine microbial species new to science, *Euhyperamoeba biospherica*, in the Biosphere 2 ocean, which was later found in very small numbers in the world ocean. The unique food chains in our ocean allowed the very primitive organism to expand its usually small presence.[12] Microbial diversity was maintained in the ocean with more than five hundred species, and food webs were self-sustaining between seeding and surveys before and after closure.[13]

The entire crew celebrated keeping our coral reef alive through the two years, the ocean being a favorite recreational place for us biospherians. Linda and I learned to snorkel during the closure. Being a landlubber and poor swimmer, I started by visiting the coral reef early Sunday mornings before visitors came on their tours. It was magic, and I learned that fins and a wet suit made up for my poor swimming skills.

We had an indicator species in the ocean—giant Pacific clams. They were beautiful, and at over two feet long, they were indeed giant, the brightness of their colors reflecting the reef's health. It was only after my time in Biosphere 2 that I saw tropical reefs: the Great Barrier Reef in Australia and the Yucatan reef in Akumal, where we collected most of our Biosphere 2 corals. Finally I learned to dive in an ocean coral reef where I didn't run into walls!

Gaie reflected on the struggles and joys of helping birth and mothering a coral reef ocean.

> There were moments when you got down to the deeper ocean where you'd sort of look around and think, where am I, because there'd be a lot of algae, and not so much of the corals. . . . not exactly like a coral reef in the wild. . . . [but] there were moments when you could be in the coral reef and you think, my God, I'm in the Caribbean, I'm in Mexico, and you would lose sight of the walls and you would be immersed in it. It would just be the most thrilling, incredible experience with the waves and everything, feeling that we really did it.[14]

Can We Really Damage the World's Ocean?

Throughout human history, the idea that humans could seriously affect or damage the ocean would have been incomprehensible, even laughable. That has changed fairly suddenly. Now, it is no longer tenable to think that the ocean can absorb whatever pollutants or chemicals are sent its way. Neither are fish populations so large that they can be regarded as virtually limitless. Relying on the size of the ocean to mitigate localized pollution no longer makes sense.

We knew the Biosphere 2 ocean was small, precious, and in great danger if we didn't actively manage it and guard it against conditions it couldn't handle. In our global biosphere, it's taken longer than it should for the wake-up call to be heard. The demand for fish in the world diet has led to 70 percent of the world's fisheries being considered overfished, with 30 percent now yielding less than 10 percent of what they once did. Overfishing is also reducing the resilience of marine ecosystems.[15]

After Biosphere 2, Gaie and Laser spearheaded the creation of the Biosphere Foundation to take the methods developed for studying Biosphere 2's coral reefs to the world's oceans. The foundation leased the Institute of Ecotechnics' sailing ship, the RV *Heraclitus*, for its Planetary Coral Reef Expedition to map coral reefs and monitor their health. At the time of Biosphere 2, no one knew the extent and location of the world's reefs to within even an order of magnitude. Using the *Heraclitus*, coral reefs were studied from 1995 to 2008 at sites in the Red Sea, the Indian Ocean, the South China Sea, and the North and South Pacific Oceans.[16] The methods used included identification of coral, fish, and invertebrates, as well as a rapid assessment technique developed by Dustan (the Biosphere Foundation's chief coral reef scientist) to monitor coral reefs using phenotypic observations.[17]

The Biosphere Foundation and the *Heraclitus* also worked with Lamont-Doherty Earth Observatory and Scripps Institution of Oceanography to drill deep cores on coral skeletons. This produced data on past climate and atmospheric conditions important for our understanding of global climate change.[18]

The health and possibly even the long-term survival of Earth's coral reefs are threatened. Sewage and fertilizer runoff in coastal waters degrade coral reefs and other marine ecosystems. Global climate change may increase ocean temperatures by another 1 to 7 degrees Fahrenheit this century, stressing marine populations. Add to these threats the acidification of the ocean through absorption of rising levels of atmospheric CO_2. This deadly trio of warming, acidification, and oxygen depletion continues to reduce ocean productivity as organisms struggle to cope with unfavorable environmental conditions. Worse, the cumulative impacts may have cascading and unpredictable impacts on the many key roles oceans play in biospheric cycles.[19]

Why should we care about the health of Earth's oceans? The question should hardly be asked since a good part of the awesome natural beauty and stunning biodiversity of the Earth is found in the ocean. "Ocean ecosystems are probably the least understood, most biologically diverse, and most undervalued of all ecosystems," concluded a global project, the Economic Valuation of Ecosystems and Biodiversity.[20]

Ocean phytoplankton produces some two-thirds of the oxygen in our atmosphere. Ninety percent of water evaporation comes from our oceans, which therefore accounts for most of the rain.[21] As Dr. Sylvia Earle, marine biologist and explorer, succinctly put it: "No blue, no green, no us."[22]

Tropical coral reefs—the rainforests of the ocean because of their high biodiversity and support of life—are gaining the attention of scientists and the public alike. Ocean acidification, which we continually counteracted in Biosphere 2, limits the ability of corals and other marine life to grow by depositing calcium carbonate. When atmospheric CO_2 reaches 450 to 500 ppm (which may occur by 2030 to 2050), coral growth will be unable to keep up with limestone dissolution. It is not feasible to try to buffer the world's oceans as we did in Biosphere 2 with additions of chemicals. The sad consensus among coral reef scientists is that many coral types will be lost, and most coral reefs will suffer major degradation unless there are dramatic reductions in greenhouse gas emissions and a slowing of global climate change.[23]

We controlled ocean temperatures to stay between 76 and 80 degrees Fahrenheit in Biosphere 2. Global warming has, since 1900, already increased world ocean temperatures by 1.3 degrees, and it could ratchet up much more if climate change is not reversed. Local hot spots form where extreme temperature rises cause the widespread death or degradation of even previously pristine coral reefs.

The Kiribati Phoenix Islands coral reefs had been lauded as exceptionally rich and healthy just a few years earlier. During the Planetary Coral Reef Foundation Expedition, the *Heraclitus* surveyed the islands in 2004 after a hot spot event. The report on the reefs' sudden devastation concluded that "remote geography did not provide a refuge from global-scale anthropogenic impacts."[24] Ocean warming has already resulted in more outbreaks of coral diseases, such as bleaching and white band disease.

Extremely damaging practices by some small, poor fishermen particularly affect coral reefs. Poison fishing uses cyanide or bleach to stun fish, especially ornamental fish for the home aquarium market. The collateral damage loss of other species and mortality of even the ones being caught is enormous. In the now devastated reefs of the Philippines, sixty-five tons of cyanide was used annually in the sea. Dynamite fishing literally pulverizes coral reefs. Each blast destroys one hundred to two hundred square feet, destroying a vital habitat and fish nursery, as well as income from ecotourism.[25]

Industrial scale, high-tech fishing causes even more damage. Bottom trawling sweeps everything up for miles. Crucial deepwater, cold-water corals are damaged, which support fish and crustacean populations. The seafloor is also smoothed, reducing habitat for marine organisms.[26] Bottom fish and shrimp net trawling results in the collateral damage (bycatch) death of 8 to 25 percent of what is caught.[27] Since fish and cold-water coral in deep waters or the sea bottom are slower to reproduce and grow, recovery from this type of overfishing will take much longer. Bycatch sadly includes some three hundred thousand whales, dolphins, and porpoises; thousands of sea turtles; large numbers of sea birds, some of which are now endangered; and large numbers of juvenile fish, imperiling future seafood harvests.[28]

Protecting Our Coral Reefs and Oceans

There is a growing international movement to create marine protected areas (MPAs), which often include restrictions on types of fishing equipment that can be used.[29] About 40 percent of all U.S. waters are in an MPA, but internationally, only 2 percent of the ocean is currently in some form of MPA. The Durban Action Plan of 2003 called for 20 to 30 percent of the ocean to be protected.[30] The member nations of the United Nations Framework Convention on Climate Change agreed in 2004 to complete the establishment of a comprehensive and ecologically representative network of MPAs and endorsed the goal of protecting and conserving at least 10 percent of all major ecological zones, including the ocean.[31]

Some see the push for a representative network of marine protected areas as the world's best hope for reversing the ocean's ecological devastation.[32] Recent research shows promise that marine protected areas can help restore marine food chains and fish populations even a good distance away by young fish that are born in the protected zones.[33]

The challenge is to come to grips with our new reality, otherwise, we face the horrifying prospect of the loss of our planet's tropical coral reefs in the coming decades as a result of our actions.[34]

A change of perspective is needed. We have to understand that our global oceans are the recipients of everything we do on land. Reducing sewage and fertilizer nutrient runoff, switching to less carbon intensive farming, industry, transportation, and energy production, and working for sustainable, regenerative limits on fishing are all part of the solution.

11

They're Mangroves, Not Mangoes!

Lost in Translation

SO YOU NEED TO CREATE a mini Everglades in the deserts of southern Arizona for a brand-new, man-made biosphere? One might anticipate that could provoke some incredulity and chaos along the way. The strangeness of the endeavor was illustrated at an Arizona state border. Arizona has stringent restrictions on plants, animals, and food in order to keep out agricultural pests and diseases. The phone calls were urgent: the trucks carrying the mango trees were being turned back. It was a classic "lost in translation" scenario. Our truckers from the deep South spoke with a heavy drawl. The border officials misheard, because it was so improbable, when told they carried mangroves. Arizona is a landlocked state. Mangoes made more sense but weren't allowed.

Yes, indeed, they were mangrove trees and other vegetation from the Florida Everglades. This wilderness biome would compress miles of estuary vegetation types. Inland freshwater wetlands are usually dozens of miles from the red mangroves on the coast. In Biosphere 2's marsh/mangrove biome, only a few hundred feet would separate them.

The plants were mostly collected in the Everglades National Park and Big Cypress National Reserve areas. They were removed at widely spaced intervals and placed in four-foot by four-foot boxes four feet deep. That ensured we also got soils, animals, microbiota, and the natural seed bank. A small portion of water was also brought from the Everglades as a yogurt culture for its aquatic life. Additionally, fish, oysters, crabs, snails, and other critters were collected. Once at the Biosphere 2 site, a greenhouse was set up with pipes allowing water circulation between the mangrove boxes.[1]

Scaling Up from Mesocosms

To test whether such an estuarine system could be successfully scaled down, Space Biospheres Ventures financed two mesocosms, which were built and researched by Walter Adey and his team at the Smithsonian Marine Sciences Laboratory. One in the basement of the Natural History Museum was mod-

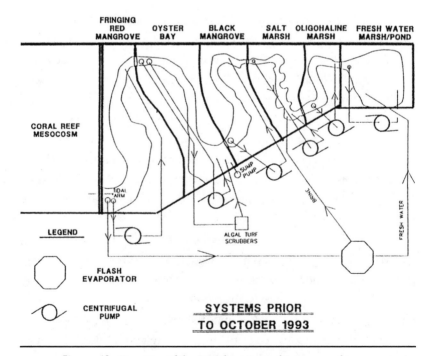

Figure 42. Zonation of the Biosphere 2 marsh/mangrove biome.

eled on the Chesapeake Bay estuary, an area under great ecological threat. The other, in a greenhouse in Washington, DC, was the prototype for the Biosphere 2 system, a smaller version of the Everglades estuary that would be made in Arizona. Both proved successful, giving confidence for the system that would be more than ten times larger.[2]

The Biosphere 2 marsh biome had six zones. Farthest from our ocean was the freshwater marsh, with wetland grasses, cattails, and trees such as cypress and willows. Next was the oligohaline (slightly salty area) with giant wetland ferns (one grows eight feet tall) and shrubs. Then came the mangrove and oyster bay zones, dominated by white mangrove trees, black mangroves, and finally red mangroves fringing the ocean. The original plan called for a tide to connect the marsh to the ocean. This connection was cut after some adverse effects were seen from the addition of lower pH, high nutrient and tannin water from the young marsh/mangrove.[3]

The Arizona Mangroves

In 1990, 208 marsh boxes, each weighing about two tons, were craned into Biosphere 2 and put in place. Like the ocean, algal turf scrubbers were used for nutrient removal.

The marsh ecosystem developed rapidly. In three years, biomass tripled from initial levels. Only 5 percent of the mangroves failed to survive, and overall growth outpaced young mangroves in the Everglades. Faster growth was perhaps due to higher CO_2 levels, lower light, and warmer winters in Biosphere 2. All the mangrove types developed similarly, shooting up from an average height of around three feet in 1990 to more than nine feet in 1993. Some of the taller mangroves were thirteen to eighteen feet tall.

Canopy development was even more robust, moving from fairly open to closed and overlapping. The shade of the tree canopy led to loss of many understory plants like in natural mangrove areas. More than two dozen fish, numerous types of shrimp, frogs, crabs, oysters, mussels, and mangrove insects were found in surveys after the two-year closure. Flowering, setting seed, and reproduction of all the mangrove species was observed, another important sign of the vitality of the system.[4]

Figure 43. The young Biosphere 2 marsh biome soon after installation in 1990.

Figure 44. Collecting leaf litter fall in the mangrove zone of our marsh, wearing a wet suit for warmth. I learned to snorkel during the two years, enjoying our coral reef and mangroves with their fish and crabs (photo by Abigail Alling).

Spending time in the marsh was a delight for the crew. It was a fun place to snorkel as well. I worked with Matt Finn, who was completing a PhD dissertation at Georgetown University comparing the Biosphere 2 and Everglades marshes, setting up litter fall and decomposition studies in the marsh (and throughout the wilderness biomes). These studies, which documented a major part of the nutrient and carbon cycling in the biomes, looked at how much leaf material was being added to the soils and sediments, and how fast they were decomposed.[5]

Linda Leigh and I also set up a schedule for observing the phenology (changes in the vegetation correlated with the seasons) of a variety of plants in all the biomes. I did the phenology studies in the marsh, which gave me another reason to enjoy time there. When the litter fall and phenology data for the marsh was compared with Everglades marsh and mangroves, they tallied pretty closely; another sign that our biome was a successful miniature replica.[6]

The marshes added variety to the life in Biosphere 2, and their rapid growth aided management of CO_2. We cut grasses and reeds in the freshwater marsh and stored them to sequester carbon and to stimulate rapid regrowth. But the marsh required little intervention by the crew. Most invasive plants can't tolerate wetlands or salty conditions. Mangroves can survive and actually grow better when they get freshwater from rain. So though they're called "salt-loving," it's just that they can *handle* the salt.

Mangroves are very hardy and self-reliant plants. They are sometimes called the "pioneers of the land" because they colonize new areas of coastal zones and help create carbon-rich peat soils as they grow. It's fascinating to study their ingenious methods for dealing with salt and aerating their root zones. Many mangroves filter out up to 90 percent of the salt of ocean water, and some mangroves excrete excess salt on to their leaves.[7] The red mangrove prop root structure is like a house on stilts. The black mangroves push up pencil-like breathing tubes from their roots, called pneumatophores.

Mangroves in the Global Biosphere

Mangroves perform all kinds of beneficial roles in tropical coastal areas. Their tangled root systems provide a protected nursery for fish, shrimp, and crabs. Coral reefs near healthy mangroves generally have much higher fish populations. These mangrove root systems also trap sediments and reduce coastal soil erosion, helping stabilize shorelines. This sediment retention also protects coral reefs from excess nutrients and pollutants. Traditionally, the people who live near mangroves use their hard and rot-proof wood for building and fuel wood, harvest medicines from mangrove plants, and use mangrove leaves as fodder. Their value is growing as a beautiful ecotourist attraction for diving and snorkeling.[8]

Mangroves also take the brunt and form a first line of defense diminishing the inland destructiveness of storms, hurricanes, and tsunamis. Six hundred feet of mangroves can lessen wave force by as much as 75 percent[9] and depth of flooding can be reduced by 5 to 30 percent.[10] The 2004 earthquake off Aceh, Indonesia, unleashed a tsunami that reached two similar Sri Lankan coastal villages. The one with intact mangroves had only a couple of deaths, while the village inland from cleared mangroves lost six thousand people.[11]

Mangroves support important wildlife, including iconic animals like alligators, crocodiles, deer, and even tigers in the Sunderbans of India and Bangladesh.[12] Many coastal birds thrive in the protection of mangrove swamps, which are important resting and feeding places for migratory birds. The two million birds using the East Asian-Australasian Flyway from the Arctic Circle to Australia and New Zealand rely on the mangroves of Oceania along their route.[13]

Defending the Mangroves

Mangroves are under assault in our planetary biosphere. By 2001, 35 percent of mangroves had been lost worldwide in just the preceding few decades. Conversion to aquaculture or shrimp farms caused more than half the losses. Industrial timber and woodchip operations, palm oil plantations, and coastal development also displace mangroves.[14]

Mangroves are being lost faster than rainforests. But until fairly recently, little attention was paid. The Aceh tsunami and other reports that showed the correlation between mangrove loss and human death and destruction during extreme weather events have helped raise awareness. New research shows that mangroves can store even more carbon than rainforests, since they create peat soils where they grow. Preservation and restoration of mangroves can be an important tool against climate change.[15]

The Ramsar Convention for the protection of global wetlands of international importance, signed by 110 countries, has led to the designation of 850 sites protecting more than 125 million acres, one-third of which contain mangroves. More than 700 marine protected areas include mangroves, many in countries with large areas of mangroves including Australia, Indonesia, and Brazil.[16] Twenty of the hundred countries with mangroves have undertaken rehabilitation efforts, attempting reforestation and replanting.[17]

Environmental groups working to protect and regenerate degraded mangroves include the Mangrove Action Project, Wetlands International, Nature Conservancy, the IUCN, and other grassroots activist groups around the world.

The mangroves were another biome that defied expectations. Transplant marsh plants from sea-level south Florida swamps to the high Arizona desert? Yet Matt Finn and his colleagues concluded from comparative studies of both: "Southwest Florida mangrove forest vegetation was successfully transplanted into a mesocosm within Biosphere 2. Dense stands of mangroves with characteristics comparable to natural Florida mangrove forests developed from the small seedlings and saplings initially installed in the mesocosm. They are good models of natural mangrove forests and can be used to learn more about mangrove ecosystem structure and function."[18]

12
Down from the Trees

Sweet Home, Our Savanna

Home, Home on the Range

HUMANS HAVE A special relationship with grasslands. We evolved there from our tree-dwelling primate ancestors. In Africa, our evolutionary jump occurred in savannas, tropical grasslands with some tree cover. Walking upright on two legs gave us a survival edge, since in the open savanna vistas we could better spot and avoid dangerous predators.[1] Environmental psychologists say our love of mowed lawns reflects historical bonding with open grasslands and the safety they offered from surprise attack.[2]

Grasslands are the in-between, the eco-transition biome. More rain shifts them to forest, less rain and they become desert. In tropical Biosphere 2, our savanna neighbored rainforest and thorn scrub/desert. Savannas thrive in quite varying climates; ones with a sharply demarcated rainy and dry seasons, long dry or wet periods, or more evenly distributed yearly rainfall.[3] Savanna plants hunker down in drought years and grow prodigiously in wet years.

The Composite Savanna

Dr. Peter Warshall, our "biome captain," decided to make the savanna a composite of the world's tropical grasslands. CSIRO, the Commonwealth Scientific and Industrial Research Organization, supplied Australian grass seed, and we sourced seed for South American, Middle Eastern, and African species. An expedition to Guyana collected termites needed for recycling organic matter that wouldn't cause damage if they escaped. Our ingenious and vigilant engineers conducted "termite taste tests" to verify they wouldn't eat through space frame sealant.[4]

Savannas have great landscape diversity: parkland, black soil (heavy, poorly drained soils that don't support trees), flat, hilly, and stone country. In wet seasons, there are flowing "billabongs," a typically dry streambed that is filled by flooding rivers seasonally and may later dry out. In the upper savanna of Biosphere 2, an Acacia tree gallery forest lined a stream that ended in a shallow pool and billabong, where water was pumped back to the

Figure 45. Dr. Peter Warshall (far right) and Linda collecting termites in Guyana, South America.

beginning. The lower savanna, reached by going down a small rise, was a sea of grasses. Our savanna was long and narrow, so the view from this rise gave us a small taste of the biome's celebrated vast, stretching horizons.

Please Disturb!

Great herds of grazing animals roam savannas and temperate grasslands. The Great Plains of North America once supported thirty to sixty million buffalo.[5] Grasses co-evolved with grazing animals and possess a unique capacity to regrow below where they are munched. Periodic heavy grazing improves the health of grasses.[6] Design meetings joked that our savanna needed a PLEASE DISTURB sign. Without herds, we couldn't put our goats to work with a thirty-foot drop from savanna cliff into the ocean. That left us biospherians to be the disturbing element.

Periodically we'd "eat" the savanna grasses with sickles. Instead of digesting the grass, we carted large bags stuffed with the dried grasses to our carbon sequester storage areas below. We moved more than a ton of biomass. For fun and to ease the work, we'd shape the bags into boat shapes that we could slide through the basement to the lung. Our savanna "bio-valve" helped manage CO_2. The cutting also kept the biome healthy. Armed with a hand sickle, cutting eight- to ten-foot-long billabong grasses that sprawled and re-rooted themselves in the mad tangle of green, the savanna sure felt far vaster.

Some grasses came to dominate as a natural ecological hierarchy established itself. Others were outcompeted in our species-packed biome. The Acacia trees in our gallery forest grew almost twenty feet, fast approaching their space frame limit. Lack of stress wood weakened them, causing some to lean over, like some of the rainforest trees.

Passion vines proved the major biodiversity threat in the savanna. Though a selected species, they luxuriated and climbed up to cover the trees. We pruned them back, enjoying small feasts of passion fruit as we worked. Once removed, the upper savanna thrived with restored sunlight in the gallery forest and grass understory. The galagos made the gallery forest their airborne highway from rainforest to thorn scrub.

Figure 46. Human (author) "grazing" the savanna.

Figure 47. Dr. Tony Burgess and Dr. Peter Warshall in the lower savanna after the end of the two-year closure, working out methods of remeasuring and mapping each grass. Dozens of grass species grew luxuriantly.

World Savannas Under Threat

Savannas worldwide—in Africa, South America, the Middle East, India, and Australia—face ecological threats causing human misery. Devastating droughts and famines ravage overpopulated, overcleared, and overgrazed savanna lands.

I've spent decades in the tropical savanna. In 1978, I helped start Birdwood Downs, the Institute of Ecotechnics' project in the Kimberley region of northwest Australia. The goal of Birdwood Downs is to develop ecological ways to regenerate the land and to plant drought-resistant grasses and pasture legumes to reverse land degradation and overgrazing.[7] The five thousand acres are typical of an enormous area of northern Australia. There is severe ecological damage from little more than a century of livestock grazing and agriculture. By tractor- and hand-clearing Birdwood Downs of invasive Acacia trees, we have restored beautiful landscapes. In the process, I fell deeply in love with the magic of the ancient outback and the resourceful bush people and Aborigines who live there. We know the difficulties and hard work needed to restore damaged savanna land.

Northern Australia's savannas face typical ecological challenges. Rainfall variability is staggering; Birdwood Downs has experienced as much as sixty inches of rain versus a low of eight inches. Our "average" rainfall of twenty-six inches means little; distribution is even more important than total rainfall. Tropical cyclones and storms can inundate with twelve inches of rainfall in a single day, but there can also be long periods between rains. Like many savannas, the Kimberley region has just one rainy season lasting four to five months. The rest of the year can be virtually rainless though temperatures remain tropical.[8]

Fire is a natural and important part of savanna ecology. Kimberley bushfires can have a thirty- to forty-mile front, burning millions of acres. Many savanna seeds only germinate when bushfires open their hard seed coats. Nature has perfected the timing; new emerging seedlings face little competition after a bushfire. Savannas often grow on heavily weathered tropical soils stripped of their nutrients.[9]

Savannas evolved with grazing animals that keep the land open for grasses because they eat, knock over, or disturb trees and tree seedlings.

Overgrazing and soil erosion mean loss of a robust grass community able to sustain fire. Lack of periodic fire, either natural or controlled burns, leads to more tree domination, often by invasive scrub trees. This was what we faced at Birdwood Downs, which was severely overgrazed, as two droving stock routes passed through the property.[10]

Early Australian settlers didn't understand the savanna. How many livestock can be supported varies wildly depending on the wet season, and lush seasons are misleading about how little forage dry years can provide. In Australia, ranches are often on an epic scale—half a million acres to a million acres. One million acres is 1,600 square miles—40 miles on each side! It takes months of roundups to reduce the number of grazing animals to meet the reality of dry or drought conditions. The cattle must be trucked thousands of miles to markets.

Much of Australia's vast savannas are moderately to severely overgrazed; desertification is widespread. At an international savanna conference I attended in Brisbane, Queensland, a South African scientist had traveled around Australia hearing about and seeing ecological devastation everywhere. He paraphrased Winston Churchill: "Never has so much been degraded so quickly by so few."

Early Australian pastoralists saw that grasses resprouted after bushfires. The "green pick" seems miraculous. In the middle of the dry season when grasses lose nutritious value, high-protein green pasture emerges from the blackened landscape. Aboriginal culture, at least fifty thousand years old, shaped the land using the "fire stick."[11] But they used patchwork, cool, small-scale fires rather than more damaging widespread hot bushfires. The hunter-gatherer Aboriginals didn't have herds of livestock. Modern pastoralists discovered through bitter experience that if they burned year after year and heavily grazed, they killed the goose that laid the golden egg.

Root reserves of pasture grasses shoot up the green pick. The more valuable fodder grasses are killed if those root reserves are depleted by heavy grazing. Then either poorer pasture plants will dominate, the area will desertify, or invasive woody scrub will take over. Cattle and sheep turnoff peaked a few decades after the Kimberley was opened in the 1880s and 1890s. Overgrazing damaged much of the best pasture land, which led to poorer pasture species dominating, soil erosion, salination, and desertification.[12]

Population Pressure and Desertification

Elsewhere, increasing population exacerbates the challenges of savanna climate and soils. Sad examples are recurring droughts and famines in the Sahel, a semidesert to grassland/savanna region south of the Sahara Desert in Africa. In the 1960s, the Sahel received above-average rainfall, and the government encouraged the movement of millions to the region. The Sahel was hit with some of the most severe droughts all across Africa in the fifty years since. Loss of trees for fuel needs, soil erosion, land degradation, and famine resulted from the deadly combination of larger populations and expanded farming and grazing.[13] Continuing climate change will cause even more devastating Sahel droughts and famines.[14]

Savanna conversion to agriculture is worldwide. Half of Brazilian savanna lands have been cleared for farming and grazing.[15] The central South American *cerrado* ranks in the top twenty-five biodiversity hotspots. It supports ten thousand plant species, eight hundred bird species, and about 150 species of reptiles, mammals, and amphibians. Only 20 percent remains undisturbed, and 2 percent is protected in the seven hundred thousand square miles of one of the world's largest savannas.[16]

The African savanna covers five million square miles—about half the continent. Protection varies widely by country. For many savanna types, only 5 percent of their area is protected. Tourism and big game hunting have helped conserve some areas of savanna from conversion to farming and livestock grazing, though some claim economic benefits do not always flow to local communities.[17]

Despite the help from ecotourism, the fates of emblematic savanna animals like elephants, lions, and tigers are uncertain. African elephants are vulnerable, and the Asian elephant is even more threatened. Lions in the wild went from a population of three hundred thousand to less than thirty thousand in the last century. They've disappeared from 80 percent of their historic African range because of fragmentation of habitat, conversion to farming, overhunting of their prey species, and killing by people protecting domestic livestock.[18] Tigers, whose habitat is both forest and savanna, now number only about 3,200 in the wild; a loss of 97 percent in the last century, including the extinction of some species like the Javan tiger.[19]

A promising but controversial approach, using grazing animals (usually considered the problem) to restore savanna and grassland ecological health, was developed by Alan Savory of the Savory Institute. Results are disputed by some scientists. The approach mimics huge migrating herds, using intensive grazing followed by a long recovery interval without animals. Called holistic management, it has been adopted by many pastoralists, including author Joel Salatin of Polyface Farm in Virginia, and often shows better results than even complete destocking in improving degraded grasslands.[20]

13

The Desert Goes Its Own Way

Fog Desert

WE KNEW OUR wilderness biomes might evolve differently than expected in the new environmental conditions of a small, tightly sealed world. So, the changes that occurred in the Biosphere 2 desert might be called a somewhat anticipated surprise.

Our lead desert designer, Dr. Tony Burgess of the Desert Laboratory in Tucson, decided early on to make a "fog desert," or a coastal desert.[1] Situated near rainforest, mangrove swamp, and ocean meant humidity would be high in the Biosphere 2 desert. Fog desert plants get a sizeable portion of their water by milking it from moist air produced by nearby bodies of water. While the Biosphere 2 desert was a mix from other coastal deserts, Baja California, Mexico, supplied most of the plants.

There, some of the strangest plants grow in North America's driest desert, nestled in a narrow peninsula between the Pacific Ocean and the Sea of Cortez. Many are endemic, found only there.[2] The "Boojum," named after a word in a Lewis Carroll poem, resembles an upside-down carrot. It forms weird landscapes with its oddly twisting top. The cardón is a columnar cactus closely related to the Saguaro cacti of the Sonoran desert, which surrounds Biosphere 2.

Figure 48. The desert biome early in the two-year closure.

Our desert zones included sand dune, salty playa, canyon, seasonal pool, and higher ground. Baja California is winter active. Most of its scant rainfall falls then, and fogs are also more frequent. We emulated this seasonal pattern, activating the desert in late fall, then stopping rainfall in early spring. The desert had a wide mix of plants, with cacti and succulents, as well as desert shrubs and grasses.

Desert plants often have distinctive aromas—a defense mechanism along with spiky thorns to deter grazing animals. The desert in Biosphere 2, with its striking vistas, was a feast for the nose. Other people knew you'd spent time in the desert by the lingering aromas you carried back.

The Desert Diverges from the Plan

The other biomes, for the most part, developed as expected, but the desert did not. Only in its drier areas did true desert cacti and succulents persist. Elsewhere, small and large shrubs, annuals, and grasses began to dominate. Added winter moisture from condensation water dripping from the space frames and lower evapotranspiration in the high humidity Biosphere 2 environment caused the shift. We let the desert evolve to more resemble Mediterranean woodland or coastal scrub chaparral ecology, augmenting this part of the original desert biodiversity by planting more Mediterranean climate shrubs and trees during the research transition after the two years.[3]

Thorn scrub neighbored the desert on two sides. The thorn scrub between lower savanna and desert was collected by a crew, including me, in the Sierra Madre Occidental near Alamos, Mexico. The lower thorn scrub, dominated by African Malagasy (Madagascar) plants, was near the freshwater marsh. This ecosystem was also winter active. The thorn scrubs were a scratchy area, and Burgess joked the thorn scrub would teach nimbleness to the crew so we could dance our way through. We kept a few vines in check, including Queen's wreath (*Antigonon leptopus*), but the thorn scrubs required only minor interventions.

Witnessing the desert self-organizing reminded us that, at a profound level, humans were not in charge of Biosphere 2. Humans were important and instrumental in keeping the technosphere functioning. We also made decisions on crop selection, rotation, and management of the agriculture. But even there, the life of the soil and the crop plants were vectors we tried to work with but couldn't control. In the wilderness areas, humans were, at best, personal assistants. Our interventions in the desert had never been very large. We controlled Bermuda grass, an uninvited species, on the sand dunes. But we mainly watched as the desert grew up and changed its nature.

During the second year of closure it became clearer that drier adapted plants no longer dominated the desert. There was little resistance from project management and crew that the best course was to step aside and let the desert change its own character. Introducing more species after the two-year closure ecologically similar to the ones flourishing accorded with project objectives. Biosphere 2 differed from botanical gardens, which work to maintain their display plants. Part of our experiment was to document the development and self-organization in our wilderness biomes. The desert demonstrated the process was working, but not as the designers anticipated.

The desert transformation reminded us of predictions prior to closure that Biosphere 2 wouldn't maintain distinct biomes at all. Some leading ecologists thought aggressive plants would dominate, turning our terrestrial

Figure 49. The desert evolves its own way. Late in the two years, shrubs and grasses dominate rather than cacti and succulents.

biomes into an amalgamation the way urban weed species dominate available niches. This did not come close to what happened. People underestimated the sophisticated technosphere providing differing environmental conditions and the ability of each biome's plants to shape their habitat.

Dissonant Voices: Let Ecology Do Its Own Thing!

H. T. Odum eloquently argued for less human interference with the wilderness biomes when he visited. His view that we should just let the biomes "do their thing" contrasted totally with our culture's emphasis on humans calling the shots, exercising as much control as possible. Similarly, Odum opposed the war on invasive species, instead arguing for accepting what nature creates, even if it includes flourishing plants from elsewhere. This perspective is counter to a certain "ecologically correct" ideology, which tries to return to a pristine, native-species-only state.

Odum even urged us to let the Biosphere 2 coral reef go: stop weeding the algae, stop working so hard to reduce nutrients. He thought coral reefs might not be compatible with the present ecological condition of Biosphere 2, but that it might after some years. He might have been right, but we ignored his suggestion. We were intent on helping our coral reef with whatever needs it had. The overall health of the ocean system and what we learned in the process justified our decision. Stewart Brand, founder of the *Whole Earth Catalog*, also argued for "going with success." We'd lost hummingbirds and finches, but a few English sparrows and a curved bill thrasher eluded our attempts to get them out before closure. He told Linda that the system was telling us through Darwinian selection what wants to reproduce, so we should let it.[4]

Biodiversity Lessons from Biosphere 2

Observations point to other challenges this grand experiment in creating synthetic ecologies revealed. The stowaway crazy ants that thrived greatly reduced our ant and pollinator diversity. They probably came in with greenhouse potting soil. Crazy ants have widely dispersed in the world and often dominate in disturbed habitats. Their dominance reflected an abundance of sap-sucking insects (homopterans), like katydids, mealy bugs, aphids, and cicadas, which the crazy ants fed on. A later crazy ant study concluded:

> Biosphere 2, a 1.28-hectare habitat island surrounded by desert, appears to be a fairly good ecological analog for a small, highly disturbed, subtropical island. As we understand more of the ecological dynamics in Biosphere 2, the results may provide important insights into the workings of simplified ecosystems that are useful for the conservation and restoration of Earth's increasingly disturbed habitats. Future research in Biosphere 2 can examine more closely the interactions among ants, homopterans, and plants and study the impact

of biological introductions, particularly introductions of natural enemies of ants and homopterans.[5]

I agree that Biosphere 2's closure experiments showed that we do not know enough as of yet to create balanced, sustaining worlds mimicking Earth's biosphere. Even sophisticated machinery, large capital investment, and what some called "the heroic efforts" of the biospherians did not prevent the necessity of injection of oxygen to continue closure (see next chapter).[6] The lessons drawn were certainly among those we hoped Biosphere 2 would teach: to appreciate what our global biosphere does in sustaining our and all life, and that we still have so much to learn about how biospheres function.

As Cohen and Tilman, the authors of the article, "Biosphere 2 and Biodiversity: The Lessons So Far," underline,

> At present there is no demonstrated alternative to maintaining the habitability of Earth. No one yet knows how to engineer systems that provide humans with the life-support services that ecosystems provide for free. Dismembering major biomes into small pieces, a consequence of widespread human activities, must be regarded with caution. Earth is the only known home that can sustain life.[7]

They incorrectly imply Biosphere 2 was intended to be an alternative to maintaining Earth's biosphere. It's surprising how frequently one encounters that assertion. Quite the contrary, the aim was understanding how crucial the services and health of our global biosphere are. Dr. William Mitsch of Ohio State University noted this in his preface to a book on Biosphere 2 research:

> The real point that resonates with me is the sheer magnitude of what it costs in money, material and energy to create enclosed healthy ecosystems . . . the ecological message of Biosphere 2 is clear—we should appreciate and try to understand the workings in the Biosphere that we have. Biosphere 2 helps us do that in many ways.[8]

Miniature biospheres will *eventually* be needed to allow open-ended habitation in space. But even that will be a progressive evolution from far simpler space life support systems. Cohen and Tilman asserted large species losses in Biosphere 2 were "unanticipated." The expectation of potentially large declines in species diversity motivated the "species-packing" strategy. They acknowledge species losses were far lower in the ocean despite its small size.

They also forget that as a long-term (one-hundred-year) experiment and laboratory facility, we anticipated periodic introductions of new species and corrective actions. Small populations are known to have genetic bottleneck problems. No one expected the first attempt at a synthetic minibiosphere to be flawless. If science already knew how to do it, we would have had less motivation to build the facility and far less opportunity to learn from its development. They note that research in a "retooled Biosphere 2 may well contribute exciting insights into the task of maintaining Biosphere 1—the Earth."[9] Fortunately, this is happening through the research conducted in the years after closure by Columbia University and now the University of Arizona.

Listening to the Voice of Nature

The desert flourished, just not in the way envisioned. Standing aside and assisting in the self-organization the desert was showing made more sense than resisting its direction. This example illustrates what's so lacking in our global biosphere. We need to listen to nature and help it, not force it to fit our preconceived notions.

Evolution is a continuing journey. Estimates of the number of species on Earth vary dramatically, from ten million to fifty million or more. We don't know for sure because the vast majority has not been identified.[10] Similarly, estimates are that more than 99 percent of the species that ever lived are now extinct. Some gave rise to successors, while others simply died out.[11] The average mammal species lifetime is one million years, though some last ten million years.[12]

To thrive ecologically and culturally and enjoy a long run on the evolutionary stage, we humans better use our intelligence and learn to change.

Maybe if all goes well, *Homo sapiens* (half-jokingly called the "human know-it-alls") will evolve to *Homo ecologicus*. Or perhaps *Homo biosphericus*: humans who understand they're part of the ecological order and act like they share the biosphere with all other life. The ancient Greek philosopher, Heraclitus, reminds us: change is the only constant. All life—species, community, ecosystem, biome—either evolves or perishes.

14

Oxygen

The Missing Element

Surprise!

AT ABOUT 9:00 P.M., January 14, 1993, I had the most amazing physiological experience of my life. Most of the crew trooped down to one of our lungs to experience what we ordinarily take for granted—oxygen, and plenty of it. Over the past sixteen months, our oxygen had been dropping.

We first discovered the oxygen decline six months after closure, when we analyzed a full set of air samples. A faulty automatic sensor failed to alert us earlier. Oxygen had fallen from 20.9 percent, like outside air, to 19 percent. The discovery triggered more frequent and detailed monitoring. We also began to search for causes.

The most immediate thought was that there was an imbalance between photosynthesis and respiration. Photosynthesis is the sunlight-powered way green plants, including ocean algae and phytoplankton, take carbon dioxide out of the air. Plants use CO_2 and water to produce sugars to build plant tissue. In the process, they release oxygen. Plants are energy storehouses, offering the sugars of their tissues as fuel to be burned in the process of respiration by humans and other animals, fungi, and aerobic microbes, to obtain metabolic energy. Respiration consumes atmospheric oxygen. Plants have their own small respiration process but mainly build tissue and produce oxygen as they grow.

Photosynthesizing organisms used that basic mechanism to change Earth's atmosphere and the course of evolution. For early anaerobic Earth life, oxygen was a deadly poison. When cyanobacteria invented photosynthesis, to later be followed by higher plants, the oxygen by-product started accumulating in the atmosphere. Then aerobes evolved with the capacity to utilize oxygen. The anaerobes retreated to places without air—deep in soils, swamps, and in the digestive tracts of animals. The last several hundred million years have been a dynamic balance between plants consuming CO_2 and producing oxygen, and microbes and animals doing exactly the opposite: consuming oxygen and producing CO_2.

Our green allies create and maintain free oxygen in our air.

Earth's atmosphere is thirty trillion times larger than Biosphere 2's, and, importantly, has so much oxygen that all the combined respiration on the planet would only bring the oxygen down one percent from 20.9 percent to 19.9 percent in more than a century if it weren't being continually replenished by photosynthesis. In Biosphere 2, we had no such luxury. The aggregate of all aerobic organisms in Biosphere 2, which included respiring microbes in thirty thousand tons of soil, would take the oxygen down about one percent a week if there hadn't been photosynthesis from plants.

But there was a mystery: Where was the oxygen going? If it was used in respiration, calculation implied we would have tens of thousands ppm of carbon dioxide in our atmosphere—which we didn't. Our calcium carbonate precipitator had captured nowhere near enough CO_2 to account for the difference.

Tracking Down the Mystery

Our oxygen research expanded to include Dr. Wallace Broecker, a famed planetary modeler, and a graduate student, Jeff Severinghaus, from Columbia University. Oxygen decline in low-light seasons exceeded that in high sunlight months when plants grow more rapidly. The high concentration of life, rapid cycling, and small reservoirs of Biosphere 2 meant far less buffer against imbalances. Earth's plants produce enough oxygen in two thousand years to replace that in our atmosphere. It only took one year for Biosphere 2's plants to replenish its atmosphere.[1]

Project management and crew agreed we'd been given a great research opportunity. Rather than immediately throwing in the towel by injecting oxygen, we'd "ride the oxygen down" in consultation with Roy, our resident physician. Human response was unknown since oxygen availability normally declines only when climbing mountains. Atmospheric pressure declines as elevations get higher; there is less air. Our bodies receive less oxygen even though oxygen concentration in the air is the same. After some time, mountain climbers adapt, the body produces more red blood cells, which carry oxygen. Time in base camps helps the climbers adapt to higher elevation.

Medical examinations during the oxygen decline showed slight or none of the expected physiological changes. This indicated that lowered atmospheric pressure, not the decline in oxygen itself, triggers human adaptation.[2] It took the special conditions of a tightly sealed biosphere to separate two factors normally inseparable.

The philosopher Socrates once said "the unexamined life is not worth living."[3] We joked that with a thousand sensors, all the biomedical data being collected, and the careful weighing and recording of every ounce of food we ate, we could update the ancient Greeks. We might change it to "the unmeasured life is not worth examining!"

Unexpected Reactions

Other reactions were fascinating. Some news media declared Biosphere a "failure," as in you said it'd be perfectly balanced; now oxygen has declined. When you inject oxygen the experiment is over and the whole project is a failure.

Meanwhile, scientists were giving quite a different perspective. Researchers in NASA's life support program telephoned. Even to those who'd been skeptical, the slow rate of oxygen decline (three-eighths of one percent *per*

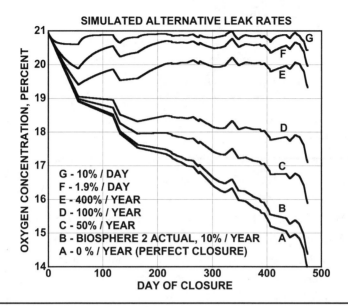

Figure 50. The impact of air leakage on detecting the oxygen decline in Biosphere 2. Line A is perfect sealing (0 percent air exchange) and is the actual decline. Line B at 10 percent leakage per year is what Biosphere 2 achieved. The other lines show how much the decline would be masked at higher leak rates (W. F. Dempster, "Tightly Closed Ecological Systems Reveal Atmospheric Subtleties—Experience from Bio-sphere 2," *Advances in Space Research* 42 (2008): 1,951–1,956). The NASA CELSS Breadboard Facility (for crop growing) at Kennedy Space Center, Florida, has a leak rate between 5 and 10 percent per day (R. M. Wheeler et al., "Crop Productivities and Radiation Use Efficiencies for Bioregenerative Life Support," *Advances in Space Research* 41 (2008): 706–713). That amount of air exchange would have produced a line between F and G in the graph.

month) showed we'd succeeded in making the structure amazingly air-tight. Otherwise air exchange would have masked the slow decline of oxygen. We might never have known oxygen was declining!

The other reaction I found quite amusing. A few said that they were aston-ished an ecological system of just a few acres would produce something completely unforeseen. It was true. I remembered everyone listing their top nightmares (concerns) at pre-closure project review committee meetings. Losing our atmosphere's oxygen was never listed.

Solving the Mystery

State-of-the-art methodologies were employed in the hunt for the missing oxygen. Isotope analysis of the more abundant carbon-12 and rarer carbon-13 helped track the pathways of our carbon. We collected new growth from plants with the three different photosynthetic pathways, C_3 grasses, C_4 trees, and CAM metabolism cacti and succulents, and soils for these carbon studies.

John Severinghaus, Jeff's father and an engineer, made a crucial suggestion: check the unsealed concrete inside Biosphere 2 as a possible carbon sink. Concrete absorbs atmospheric carbon dioxide through a process called carbonation. Inside, there were large areas of bare concrete in floors, walls, and structural columns. So, we analyzed cores of concrete that had been poured inside and outside at the same time. The inside concrete had ten times the amount of carbonation. Far higher levels of inside CO_2 caused more carbonation, which also explained why oxygen loss was greater during the winter time: less oxygen production by plants led to far higher levels of CO_2.

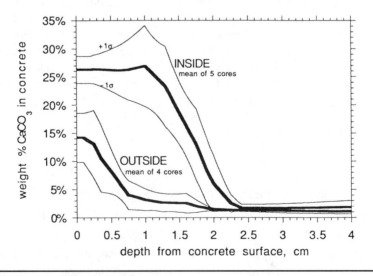

Figure 51. Data on the stronger carbonation of concrete inside Biosphere 2. Concrete was tested from batches poured at the same time inside and outside the structure. (J. P. Severinghaus et al., "Oxygen Loss in Biosphere 2," *Eos, Transactions American Geophysical Union* 75 (1994): 33–37.)

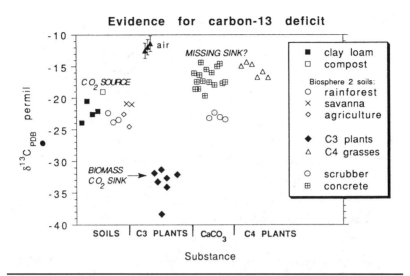

Figure 52. Part of the research discovering where the oxygen went in Biosphere 2 by tracking the carbon-13 isotopes. (J. P. Severinghaus et al., "Oxygen Loss in Biosphere 2," *Eos, Transactions American Geophysical Union* 75 (1994): 33–37.)

The conclusion was clear. The culprit was the imbalance between photosynthesis and respiration by the microbial life of the thirty thousand tons of soil inside Biosphere 2. The missing oxygen ended up in the concrete by the respiration pathway through CO_2, concealing the simple fact of the imbalance.

Project designers had initially made a difficult choice. Overstock at first, or limit resources and risk slowing growth. The balancing act was that we needed a rich, organic soil to promote rapid development of plants and trees in the biomes and crop production in the agricultural area. For the short term, we expected some imbalance between photosynthesis and respiration. As plant biomass increased, we hoped to reach more stable atmospheric dynamics. About five years after the first closure, Biosphere 2 soils, particularly farm soils rich in organic materials, dropped from an initial C:N ratio of 16:1 to a C:N ratio of 12:1. This is typical of productive agricultural soils and within the stable range for carbon and nitrogen. Thus, in just a few years, Biosphere 2 had far less soil oxidation of carbon.[4]

Adaption or Hibernation?

Half the crew started having symptoms associated with lowered availability of oxygen such as sleep apnea, as we fell below 16 percent oxygen. This condition makes people suddenly wake up from sleep because the body senses a lack of vital oxygen. To alleviate these symptoms, we ran lines from the analytic lab oxygen concentrator to four of the crew rooms. They put on oxygen breathing tubes to counteract sleep apnea.

Lowered oxygen can impair rational thought and judgment; therefore, our on-call medical staff was told, if necessary, they could override Roy Walford's decisions. When Roy discovered he couldn't add up a line of numbers, he

Figure 53. The crew relaxes and tries to rest during pruning of the savanna. Note Taber resting with his sickle still in his gloved hands. Calorie and oxygen restrictions taught our bodies to be precise and not waste any effort.

asked the consultants to make the call. Everyone agreed it was time after riding the oxygen decline for sixteen months to about 14.5 percent oxygen. The biospherians had agreed in the name of science to ever more biomedical testing. We rode stationary bikes with instruments attached to our bodies, we blew into special devices, and we even drank spiked isotope drinks so our urine could be further analyzed.[5] To avoid potentially more serious health issues, engineers would inject enough pure oxygen to bring levels up to 19 percent.

We went into the west lung where the oxygen was being held overnight so accurate measurement could be made before the oxygen mixed into the atmosphere. It was a cold 50 degrees Fahrenheit since the pure oxygen had arrived in refrigerated trucks. The response to 26 percent oxygen was near immediate. Linda remembers:

> It was a sense of well-being to breathe deeply and not feel like I needed another breath almost immediately. Barely noticing the people outside looking at us through the lung window, I got a totally unexpected and sudden impulse to run around the lung for no conscious reason, just an impulse which drove my legs. I felt like a born-again breather praising oxygen to my formerly breathless companions.[6]

We were all madly laughing and running. Then it struck me, I haven't heard the sound of running feet in months. Friends said watching us biospherians work was like seeing a dance in slightly slow motion. With the restricted-calorie diet and lowered oxygen, people were economical in their movements. There simply wasn't extra energy to burn.

Roy and his colleagues studied our medical and physiological data. They theorized calorie restriction made our bodies reluctant to make more red blood cells. Their grand theory was that we were the first humans starting to display symptoms of hibernation to deal with both oxygen deprivation and a reduced-calorie diet![7]

After the jubilation in our oxygen-rich lungs, we started walking up a flight of stairs to the human habitat, which was another astonishing experience. With each breath and step, as we entered an atmosphere at 14.5 percent oxygen, I could feel my breathing becoming shallower, my movements slower. Once again, speaking a long sentence forced us to take a breath to

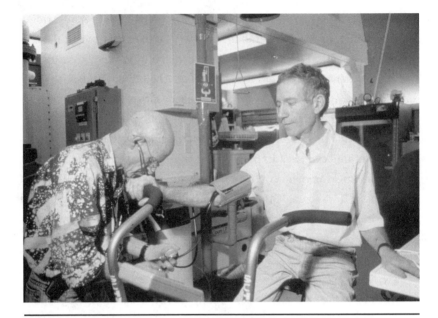

Figure 54. Riding an exercise bike for science. Our already-full schedule of medical checks and research ramped up with the oxygen depletion work. For one study, we all drank water with special isotopes to permit detailed investigation of our physiology and metabolism. (C. Weyer et al., "Energy Metabolism after 2 Years of Energy Restriction: The Biosphere 2 Experiment," *The American Journal of Clinical Nutrition* 72, no. 4 (2000): 946–953. Photo by Abigail Alling.)

make it to the end. Gone was the teenager I'd felt in that oxygen super-charged lung. When I reached my room, I felt older than my forty-five years.

What We Take for Granted

The oxygen loss exemplified what we could learn from Biosphere 2. Solving its mystery revealed so much about critical balancing issues between our soils, plant life, human physiology, and atmosphere. From what it taught us, future biospheric laboratories can be much better designed. Our consultants had foreseen even if the early biosphere achieved reasonable balances, later it might be necessary to bring in additional carbon and other nutrients when initial soil reserves were gone. Despite Biosphere 2's complexity, causative mechanisms could be discovered. It only took a year to nail down the mysterious oxygen sink.

Contrast that with the global carbon budget. For thirty years, the "missing carbon" has vexed Earth system scientists. It's not a small amount of unaccounted carbon. Humans produce seven to eight billion tons of CO_2 annually. Where one to two billion tons (15 to 25 percent of the total) goes is unknown.[8] Before closure, I had attended conferences where ocean scientists and terrestrial ecologists accused each other of not having good enough data to find the missing sink. Currently, it's thought northern boreal forest soils are the missing sink. Research is still far from definitive, with some data suggesting the sink is divided between boreal and tropical forests.[9]

Other appraisals pointed out strikingly analogous behavior between what was happening in Biosphere 2 and global ecology. The grand experiment in ecological self-organization was underway with so many lessons to learn.

Examination of the ecosystem within Biosphere 2, after 26 months of self-organization . . . [showed the] system appeared to be reinforcing the species that collect more energy . . . species diversity of plants was approaching normal biodiversity . . . the observed successional trend (carbon dioxide absorption by carbonates and high net production of "weed species vegetation") if allowed to continue, was in a direction that would eventually generate enough gross production to match respiration of the soil, which was gradually declining. Thus the self-organizational development of a human life support was successfully underway . . . the smaller, faster Biosphere 2 is a good model for studying the biogeochemical dynamics of our Earth.[10]

The metabolism in Biosphere 2 turned out to be a good analog for that of the whole planet Earth . . . The study of Biosphere 2 provides the important insight that long range oxygen levels on Earth are partly controlled by the cycles of calcium and carbonate, and that only by considering the interaction of these biogeochemical cycles can we gain a realistic understanding of planet homeostasis and feedback mechanisms. Biosphere 2 because of its analogous properties to the Earth provides the opportunity to investigate the potential implications for the global carbon and oxygen budgets that may result from climate forcing and a [higher] carbon atmosphere.[11]

As countdown to re-entry approached, I took many trips around Biosphere 2, daytime and nighttime. I wanted to imprint the deep experience of being so connected with all our living systems inside—from farm crops to wilderness biomes. There were considerable mental and emotional reactions as well as actual tree-hugging. We, all of Biosphere 2's life, had been on an epic journey together.

I took time to hug and thank a few concrete pillars in the technosphere basement. They did their job too. If they hadn't absorbed all that carbon dioxide, we might not have maintained acceptable levels. The concrete saved our coral reef by preventing pH from getting too low and helped plant growth by keeping CO_2 from reaching dangerous levels. They helped us succeed—which was an open question until the end—in being able to live inside Biosphere 2 for two years.

They had their own lesson to teach us. Everything in a closed ecological system plays a role, even ones we don't anticipate. Ecology's first law is everything is connected to everything else. Let's extend that thought to every part of our technosphere as well.

Biosphere 2 taught us to never take anything for granted, including what we hardly ever think about: the oxygen that keeps us alive we owe to our biosphere.

15
Humans

The Most Unstable Element

Have Courage!

OUR RUSSIAN COLLEAGUES in closed ecological systems had decades of practical experience. While NASA loves controllable technology and seems to always want to limit life in space life support, Josef Gitelson told us what they'd learned from Bios-3: "Trust life, it's reliable; nature has perfected it over billions of years of evolution. Count on technology to break down—it's not a question of if, but when."

Just before John Allen entered the Biosphere 2 Test Module for the first human closure experiment, Dr. Yevgeny Shepelev sent us a message. "Congratulations on your historic step. But remember, humans are the most unstable element in the ecosystem. Have courage!"

We suspected group dynamics might be one of the more challenging aspects of Biosphere 2. Turns out, we were right.

"Irrational Antagonism"

Cabin fever leads to the buildup of tensions when people are in close confinement. This happens even among good friends who respect one another,

Figure 55. Best of friends. Seven of the crew right after entering. Standing, from left: Sally, Gaie, Taber, Linda, author; kneeling, from left: Roy, Jane (photo by Robert Rio Hahn).

and it can get worse with prolonged isolation. Psychologists call it "irrational antagonism."[1] In expedition circles, it's called "explorer's cholera." Admiral Byrd, the great polar region explorer, wrote: "I knew of one who could not eat unless he could find a place in the mess hall out of view of a [person] who solemnly chewed twenty-eight times before swallowing. In a polar camp, little things like *that* have the power to drive even disciplined men to the brink of insanity."[2]

For us biospherians, though our world was diverse and beautiful, it was limited. We only had close physical contact with seven other people for 731 days, 2,193 meals. Pretty quickly, everyone figured out tics, habits, and how to push the others' buttons, not unlike dysfunctional family dinners!

Gaie and I wrote a book, as yet unpublished, during closure. One chapter, we titled "The Breakfast from Hell." Crew members deliberately provoked each other, at first earnestly and maliciously in attack and counterattack. Finally, in fun, we realized what was going on. Pushing one another's buttons was a sport that could go on endlessly—and sometimes it seemed it did—since we knew each other incredibly well.

There can be a positive side. You can have the most profound or distressing experiences (or anything in between) since you have to draw on your own resources. Jane reflected, "We're so used in the [W]estern world to being constantly bombarded by events, by all kinds of things happening, by running here and running there. And when you're in an isolated environment most of that goes away. You're left with yourself. You're left with your own brain, your own mind."[3]

This can be an opportunity.

You have seven mirrors to face each morning, lunch, and dinner, reflecting back an aspect of yourself you want to change. If your crewmates found it unpleasant, you had to live with yourself. Living in Biosphere 2 was a perfect place to work on changing character traits after being able to fully see what got under others' skin. It was two years of intensive group therapy "marathons": extended periods of time with the same people. The alchemy of the two years resulted in some very deep restructuring for me, and I'd guess for all of us. In addition to personal work on myself, the extraordinary circumstances of our life and the consciousness it catalyzed must have contributed.

Group Dynamics: Theory and Practice

Biospherian training included both group dynamics theory and working with small groups at remote sites. We studied the approach of W. R. Bion, a pioneering English psychologist who discovered basic small group mechanisms. During World War II, he worked with groups of British Air Force pilots with PTSD. He discerned two contrasting modes that govern groups. A "task group" remembers their objectives and intelligently uses available time and resources to accomplish goals. The other mode is "group animal."

Usually unconscious, group animal manifests as "kill the leader," "dependency," "pairing," or "fight or flight" group behavior. A group that remains aware of these mechanisms tends to keep on task.[4]

But there was no preparing for an experience no group of people had ever tried. Once the airlock shut behind us, we were only eight.

We came in as friends. Everyone was committed to the experiment's success. Despite heavy workloads, people voluntarily helped with additional sample and data collection and undertook new projects with outside scientists.

We faced numerous stresses apart from physical isolation. These included calorie-restricted diets, lowered oxygen, and media that either lionized or trashed the project. We fell into the long-simmering controversy between analytic and holistic systems science. H. T. Odum commented, "Some journalists crucified the management in the public press, treating the project as if was an Olympic contest to see how much could be done without opening the doors."[5]

Figure 56. Roy gives a talk at the Third International Conference on Closed Ecological Systems and Biospherics, Biosphere 2, April 1992.

We used every type of communication available. We telephoned people and met friends and colleagues at the windows. We talked through two-way radios to those inside and outside, video conferenced, and used email through the primitive Internet. We gave talks and participated via video in our command room with workshops and conferences. Biosphere 2 hosted the Third International Conference on Closed Ecological Systems and Biospherics in April 1992. I organized the participants by phoning them. That meeting devoted one day to carbon studies with key global climate change scientists. We were beamed into NASA centers, the International Space University summer session in Japan, the Explorers Club annual dinners and space conferences like The Case for Mars in Boulder, Colorado. We wrote scientific papers and popular press articles.

Interacting with school groups was always a morale booster for the crew. Even when the project was attacked in the press, seeing the enthusiasm of hundreds of thousands of visitors made us feel the optimistic and dramatic story of Biosphere 2 was reaching people.

Media and the Art of the Sound Bite

We only received media training a few months before closure, not anticipating the worldwide interest in the experiment. Project managers called in Carole Hemingway and Fred Harris of the Hemingway Media Group. Thank goodness they did!

It's hilarious watching the earliest videos of the crew learning how to do media interviews. We clearly thought it an incredible imposition to stop working. Our body language said, "We have lots of stuff to do to prepare ourselves and the facility for closure. And you want us to sit and be videotaped responding to questions?"

Media training was essential. Otherwise we'd be complete lambs to the media slaughter, not understanding what we were walking into. The training was also extraordinarily effective psychotherapy and objective mirrors of our behavior. On video playback, it's clear no matter what we thought we were doing: our body language, subtle emotional states, and sloppiness of dress or formulations were painfully obvious. It's embarrassing and a revelation to watch yourself.

Carole taught us the art of sound bites. They've declined from more than forty seconds in the 1960s to nine seconds.[6] Say anything longer and it'll be edited out. Try formulating anything of substance in nine seconds! It takes work and focus to get it right. Speak slowly and distinctly to seem more assured and knowledgeable. As a fast-talking ex-New York City kid, Carole had me watch the movie *Patton* to underscore this point.

We made a vow. We wouldn't trivialize our experience in Biosphere 2 by talking about our sex lives or who and what we miss. It's weird how some American media were fascinated whether there would be a baby born inside. I recall one testy answer: "We are modern women and can practice birth control. We have plenty of work as it is."

The *Arizona Republic* noted virtually every news interview included questions about sex, whether we paired up like passengers on Noah's Ark. They repeated two of our evasive answers. I said, "People are people. Everything you might expect to happen with people has happened in here." Sally responded, "Everyone was perfectly free in here to make relationships how they wish."[7] We didn't want to dwell on the mundane, personal level. Of course we would miss family, friends, going out for a meal or movie.

When eight was decided to be a good size crew, project leaders made a statement about gender equality; half would be women and half men. Being single, married, or partnered was not a selection factor. There were four people without an inside relationship and four with one: Gaie and Laser, Taber and Jane. Since our privacy was important, we didn't speak about sex—to the media or to one another! Jane wrote that having Taber as a partner was a great source of joy and stability.[8]

For me, not having a sexual relationship reinforced my feeling that Biosphere 2 was my Zen retreat. On trips to Japan for space conferences, I'd dreamt about giving everything up and entering a Zen monastery. I loved that Biosphere 2 was a simplified world and life, with definite responsibilities and routines, allowing more time for thought and reflection. Thoreau growing his beans at Walden Pond had urged: "Simplify, simplify, simplify!" I spent two years as a happy eco-monk.

Well, maybe there were unfulfilled longings. One Valentine's Day, feeling lonely, I got a zap when my eyes met those of a woman touring outside the facility. That energy boost fueled my entire day's work! I also identified with our lone stowaway curved bill thrasher. I'd see him perched on a tree in early evening looking out at the desert outside the glass, filling our world with his lovely song. Serenading a lost love outside or hoping to entice a mate? Didn't he know our sealed structure meant his song didn't leave this little world? Maybe he sang because he was at peace with the world.

With more time for reflection, I kept an extensive journal, usually filled out late at night. For the first time in my life I spent a good portion of my free hours writing. With calories low and oxygen more so, all of us unconsciously conserved our physical efforts.

Media discipline, especially for network TV, meant coming armed with six or seven sound bites about what was happening in Biosphere 2. There was the sport of "question turn around." No matter what they ask, turn the question around (gracefully) to get back to one of your sound bites.

This press and media interest consumed a surprising amount of our time. Jane was our hairdresser and makeup person for TV interviews. Laser did the technical setup with cameras and microphones. Often in those early days of global communications, there'd be hours of a series of uplinks with media around the world, with crew taking turns being interviewed.

Figure 57. Gaie speaking to the media via video from the Biosphere 2 command room. It was stressful knowing the interview and media coverage might reach millions of people, but we were buoyed by the intense interest and support of people around the world.

All the media attention offered golden opportunities for public education and communicating the reality and challenges we were experiencing. As our communications officer, I fought in the early months for the project to basically share everything in real time. Since there were issues of intellectual property for a privately funded venture, there had been restrictions on release of data and photographs. When we started doing weekly and monthly reports, our relations with the press improved.

The project managers and biospherians had no illusion everything would run smoothly from the outset. We expected with some certainty we'd have to make changes to keep Biosphere 2 functioning. It's embarrassing to recall press releases before closure announcing: "No air, water, or material will cross the airtight boundary of space frame glass between Biosphere 2 and the surrounding biosphere of Earth."

The project made many mistakes. It's understandable when you consider how new and difficult the endeavor was. In hindsight, we should have shared the top unknowns and concerns of our designers and consultants, including more than a few nightmare scenarios. This would have underscored that Biosphere 2 was an experimental facility. If we thought everything would work perfectly, why bother to build it? I would also have loved if part of our technosphere basement was viewable by outside visitors to appreciate how much engineering equipment was needed to keep our little world functioning.

La Vida Ecologica

We strove to live as complete, satisfying, and "normal" a life as possible. We were not deprived by being ecological! We laughed at letters and journalists' assumptions that we might be cleaning our clothes by washing them in one of our streams—we had washing machines! It was a pleasure as well as responsibility to live with and take care of our living world. Our primary job was to keep the machinery going. Sometimes we intervened in the biomes as Team Biodiversity.

The breakdown of crew time shows the varied lives we led. Farming took 25 percent of our labor, communication (including writing reports and papers) took 19 percent, food preparation required 12 percent, and taking care of domestic animals took 9 percent. The terrestrial biomes and research took 6 percent each, marine systems 5 percent, repair and maintenance 4 percent each, analytic and medical labs including research took 3 percent each. Finally, sample export and media interviews required 2 percent each.[9]

We had a strange public privacy. But visitors' enthusiasm always boosted morale. It took getting used to being watched while we worked. This was my journal entry three days after we entered:

> We harvested two rice paddies Friday morning—very good yields—but suddenly as I was coming to pick up more sheaths of rice, I noticed 40–50 school kids looking through the glass. People now come all around Biosphere 2. I confess it takes getting used to. All of a sudden after working by yourself, you look up and there's a group, excited to see you, trying to get your attention, cameras clicking. I'm getting over my shyness, learning to smile and wave and continue what I'm doing. It's instructive, makes me pay more attention to the precision, the grace with which I'm working.

We all felt it was part of our educational outreach for people to see us growing food, taking care of a sewage treatment constructed wetland, tending to the coral reef, taking care of our wilderness biomes. We were a living exhibit: this is how people live with a biosphere, and we treated it with tender loving care.

The Intrusion of a Power Struggle and Outside Politics

Absent from Biosphere 2 were money, packaging, and trash. But we were not immune to politics and power struggles. I can't separate the ups and downs of our group dynamics without wondering how much our problems were exacerbated by an outside struggle over the management and direction of the project. Early on, although we were all friends, some social differentiation started to occur. To some extent, two informal groups began to form. That's normal; in any group some people get along better and like to hang out with one another.

Figure 58. The "Bio Band," with Taber on drums and Jane on vocals and keyboard. Roy sometimes joined with his electronic sax. Their recordings incorporated the inside sounds of nature and our technosphere (photo by Abigail Alling).

The outside power struggle polarized the inside team. Tensions increased markedly. The more compatible groupings of four became aligned on opposite sides. Sally, Gaie, Laser, and I backed the project's leaders and original intent. Roy, Linda, Jane, and Taber favored changes.

The core issue was whether the primary goal was to work on maintaining as self-sufficient a closure as possible and improving the facility, including trying to produce all our food. The alternative was to lessen workloads by sending in food to give more time for research. This first closure was called the "Shake-Down Mission."

A faction of the project's Scientific Advisory Committee (SAC) went past making research recommendations and got involved in recommending changes in project management. That made the issue far more incendiary. Now looking back on all this, it seems a bit crazy that this was the cause of so much struggle, internal division, and, at times, bitter antagonism. But, the prospect of a change of both management and project direction gave license for other grievances to surface, such as who should be in charge of what aspect of running Biosphere 2 among the eight of us.[10] When the chain of command is challenged, powerful emotions can be unleashed.

The power struggle launched a charged period when our group dynamics and interpersonal relations were the hardest. We had followed a weekly format developed at the ecotechnic projects. It included three special evenings. Tuesdays were cultural evenings for a movie (new releases were electronically sent in), music, book discussions, or to look at insects and flowers under a microscope. Thursday evenings, we'd discuss psychologically interesting topics or readings from different spiritual traditions, exercises in nonlinear thinking. Sundays were our celebratory dinners, with home-brewed wine and toasts, then personal reflection speeches afterward. Originally, Saturday mornings were for acting and movement workouts. They were also for

Figure 59. The Interbiospheric Arts Festivals. Via two-way video we linked with outside artists, musicians, filmmakers, and poets, presenting our work and listening to and seeing theirs. The crew was motivated to find artistic ways of exploring and communicating the experience of living in Biosphere 2.

developing a theater presentation, either an existing work or one developed through improvisation.

Theater: Psychodrama Therapy

Theater is an excellent way to work on group dynamics and morale. Putting your daily life into dramatic form helps work out personality conflicts and draws people together in a deep way. For many years at our Kimberley, Australia, projects, we improvised annual outback comedies, poking fun at ourselves and life on the frontier. We performed it to appreciative audiences at our town's local festival, nearby towns, Aboriginal communities, or at our own homesteads.

Before closure, the then biospherian crew (I was a late substitution) developed a hilarious theater piece called *The Wrong Stuff* working with IE director and theatrical genius Kathelin Gray. It was eerily accurate in some ways, containing scenes about frictions between the inside crew and mission control. One refrain went: "cut the cable, cut the crap." There were also petty acts of food thievery, complaints about endless farm work, and power struggles over decision-making and cabals.

But acting and movement workouts were dropped early in the two years, with the argument that we had too much to do and fear of misinterpretation by a critical media. During the height of the outside power struggle, there began to be evenings when four people (the "dissident" group) took their meals to eat apart from the rest of the group. In addition, individuals would either elect to take part in other weekly events or boycott toasts and speeches, deflating what were supposed to be Sunday celebrations. There were no questions asked, as all these activities were voluntary. But they were troubling as reflections of discord.

Dark Days

These were the dark days when, as toastmaster, I had to disallow a toast "to the traitors" (meaning those that backed a change of management). People became more reluctant to share their inner life through the weekly speeches. There were also two incidents of spitting (which I only learned about later). It was dicey, as clearly a few of the crew disliked one or two of the key management leaders of the project on the outside and wanted them gone.

Linda acknowledged years later: "The split was contentious. When I look back at my behavior during that time I feel pretty shocked. It was really awful. Where I would be so cold to the people in the other group, walking by them and not even looking at them. And, you know, you're in a closed system—there're only eight people—that's pretty awful."[11]

Roy Walford wrote that there was "a split between those who strongly supported and those who strongly resented interference from Mission Control."[12] But all this was to be expected. Displacement of anger and tensions between the crew and mission control are well-known phenomena in space crews.[13]

One crew of cosmonauts cut Earth communications for a day. Mission control never asked why, but was relieved the next day when the crew started talking to them again.[14] Oleg Gazenko had served as the unofficial elder friend and psychiatrist to generations of Soviet cosmonauts. He told us some of their conflicts were only revealed many years after they returned to Earth. That's part of the Alpha personality: show no weakness or emotion. I recall reading in a Russian book about its Salyut space station that one cosmonaut wrote that living in such close quarters was a perfect recipe for homicide!

Periodically, we were reminded by mission control that we were volunteers. At any time, no questions asked, we could leave through an airlock. I don't think any of us were tempted. We had worked so hard to get there, and it was too interesting and deeply satisfying staying inside.

But the strains of living in a small world with a limited number of people were real. Reviewing my Biosphere 2 journal though, I'm struck by how much group morale and tension fluctuated up and down. There were periods of greater friction and conflict, and periods of more cohesion and higher group morale throughout the two years. We certainly worked at it. We even did a group rereading, chapter by chapter, of Bion's book as one of our Thursday evening series. Afterward, we discussed in detail how our overall task was progressing and tendencies for "group animal" modes to manifest. One evening we passed a "talking stick" and allowed everyone to have their say

Figure 60. Snow covers the landscape around Biosphere 2. These were times when the crew yearned to go outside to play in the snow. The contrast with our tropical, T-shirt-and-shorts environment was vivid when seeing people outside in heavy overcoats. But we had the feeling that there was always more to explore and learn inside.

uninterrupted. I think these all helped. Even if we didn't resolve underlying tensions and disagreement, that we laid them out meant that we remained conscious of them.

Three months before leaving, I was asked if not only ecological cycles but human and social ones are sped up in a closed system. I recorded my response in my journal: "From our experience I think so. I have been to Heaven and Hell a myriad of times in here, been enlightened, been profoundly discouraged, you name it, I've experienced it." Biosphere 2 was indeed a cyclotron of the life sciences.

Rebecca Reider spoke to all eight of us when she expanded her Harvard history of science thesis into a book. She commented in amazement: "I was struck by what different mental universes the two foursomes inhabited. They had shared their physical universes as intimately as eight people could . . . eating the same food, breathing the same air, facing all the same situations for two years. Yet listening to them describe their relationships, it sounded as though they might as well have been living in different biospheres."[15]

Frictions began early on. Roy complained just weeks into closure that working in the farm didn't allow him enough time for medical research. My journal records Laser's angry response: "We're going to make a new way of life, and you want to divide us into farmers and scientists!"

Though Rusty Schweickart had warned us about avoiding the "us vs. them" split that occurs in isolated groups, it happened nonetheless. Jane even capitalized the words in her memoir. "Us" were the good guys "on the side of science" and "Them" (Gaie, Sally, Laser, and I) were "loyalists" to management, and maybe less pro-science; though we'd argue we were pro-science, including integrative and holistic science.

Pondering these dynamics, as I still do, I'm struck by the realization that the crew divide was not primarily between the crew. At its core, it was the classic divide between the explorers in the field and outside project management, astronauts and mission control. Some researchers note that animosity toward mission control usually serves to further unite the crew.[16] But in our case, it didn't.

Crew attitudes toward mission control totally varied depending on which "mental universe" one inhabited. I can recall vehement discussions with Roy who asserted that we biospherians were more micromanaged than astronauts. Having spoken with more than a few cosmonauts and astronauts, I couldn't believe this contention. Astronauts generally have their days lined out virtually minute to minute. Even on missions like Skylab when the astronauts received more leeway to organize their workdays, it was because they'd be better able to accomplish the mission objectives. From my perspective, we arranged our own days and tasks in crew meetings. We made innumerable operational decisions inside, including changes to our diet and crop selection and rotation.

Early on, we banned all meetings and unnecessary work on Sundays (aside from animal feeding) so we'd have a complete day off. We also voted, as a morale booster, to take all holidays our outside staff were celebrating and invented holidays in months without them.

When I reread my journal recently, I was struck that group dynamics and crew morale were not at all linear—tensions waxed and waned. So I read with astonishment when Jane asserted in her memoir, "that's the way it was for the last 14 months. We never looked each other in the face again."[17] Really? My journal and memories say otherwise.

Even during periods when others remembered only ill tempers, there were breaks and periods of joy. For example, the following was my journal entry for June 10, 1993:

"Well, well, well. Extraordinarily good chemistry at our weekly crew meeting. Lots of humor, free flow of conversation and jokes. Decided spontaneously to celebrate the Captain's Birthday tomorrow as a holiday, in the tradition of taking a holiday for the Queen's Birthday as they do in the U.K. and Australia. We'll tell the outside staff it's a day for "catching up" and writing papers etc. Roy sort of started it by semi-seriously saying that he'd been badly bitten by our local fire-ants near Tiger Pond [under the rainforest mountain waterfall], and if he seemed delirious or spacey, that was the reason. From which ensued endless jokes about that being the reason for everything under the sun."

When we wanted to initiate new research programs with outside scientists, it was never vetoed by mission control. They knew we had full workloads but agreed with our desire to increase research.

Roy, as a full-tenured professor at UCLA, was used to calling his own shots and running his research lab as he wished. But Biosphere 2 was not built for eight biospherians to do whatever they felt like inside. There were sizeable financial investments, and the project was ultimately directed by top management. I told Reider, "Imagine NASA astronauts deciding they were tired of mission control instructions and the overall plan for their space mission!" Gaie agreed. "Anytime you're in an organization . . . whether you like it or not, there's a boss."[18]

Of course, this dissension was empowered by the SAC faction trying to change project management. There was now a challenge to the way Biosphere 2 was built and operated. The SAC issued a report in July 1992, ten months into the closure. They called the project an "act of vision and courage" and noted "Biosphere 2 is already providing unexpected scientific results not possible through other means . . . Biosphere 2 will make important scientific contributions in the fields of biogeochemical cycling, the ecology of closed ecological systems, and restoration ecology."[19]

The SAC also gave a series of recommendations to improve the science program. Almost all were implemented, such as the import and export of samples and equipment, the hiring of Dr. John Corliss as a separate director of research, and a call to more formally establish the research program. Roy and Gaie would soon produce a program with more than sixty specific research projects. But not all recommendations were liked when carried out. The analytic lab equipment was sent out so outside staff could operate it to reduce our workload. Taber was unhappy his beloved lab was emptied, and we were all frustrated by the time it took for analyses to begin again outside.

It's hard to convey the stress we felt inside. Negative media fixated on whether we were doing "real science," the allegations of "cheating" because of Jane leaving and returning with a duffel bag, and the not-particularly-secret carbon dioxide scrubber. At times, it felt like a pressure cooker. None of us were used to being so high profile and to then being excoriated in coverage that reached millions. I remember *Newsweek* claiming the ocean was dying and that crew went out daily in boats retrieving dead fish. Pure invention, and we laughed. Didn't they know we didn't have that many fish to start with?

Crew comments reflected our feeling of disconnect. Gaie said, "Can anyone just say we're just doing okay, can someone give us a pat on the back, please? Is there anyone out there, except for the immediate people working on this project that are all under attack, that can extend a hand of friendship, and just appreciate the glory of it all?" Jane noted, "It was insane that we had that much animosity from the media, because we were working our guts out inside on a really very worthy, worthwhile venture." I concurred: "The press decided that if Biosphere 2 isn't perfect, it's a failure."[20]

Under that stress, it's little wonder that some of the crew decided to align with what they perceived as the highly credentialed SAC, or the faction of it wanting to oust management. Voices like H. T. Odum were rarely quoted in the media: "Good ecological engineering involves incremental changes to fit technological operation to the self-organizing biota. The management process during 1992–1993 using data to develop theory, test it with simulation, and apply corrective actions was in the best scientific tradition."[21] Before and during the two years, distinguished scientists came to give lectures and learn about Biosphere 2.

There were also deeply personal issues at play. Jane would honestly write later, referring to John Allen: "[F]or 10 years I had believed him infallible, held him in such high regard that no one could have lived up to the god-like image. When he showed himself to be chipped, cracked, less than perfect, humanly flawed, my adoration turned to hate."[22]

And more mistakes were undoubtedly made. Management outside must have also felt under great pressure to bunker down and adopt a siege mentality. They felt undermined by part of the crew and limited their communications with outside scientists for fear of them adding fuel to a takeover. These dynamics engendered bad feeling inside and outside Biosphere 2.

I tried unsuccessfully to get Roy on our side, realigned with the original vision of Biosphere 2. We argued about different approaches to science, the legitimacy of non-hypothesis-driven research. He was the most credentialed of the crew and the most influential of Jane's "Us." It came later, a few years after our closure, but Roy deeply rethought these narrower definitions of science that are prevalent in most of academia.

When Roy wrote his 1999 paper, "Biosphere 2's Voyage of Discovery: The Serendipity from Inside," he compared the project to Darwin's voyage on the *Beagle*, undertaken without any specific expectation of what he'd find. Roy endorsed the importance of "integrative science," which "attempts to

Figure 61. Body language and facial expressions say it all. The crew gathers at our video conference table to discuss group tensions with mission control. All eight were there; two of the crew were cut off in this photo.

deal with nature on a larger front, where knowledge of the system remains incomplete, surprise is inevitable, prediction is hazardous, and emergent properties may be encountered, springing from the internal nonlinear dynamics of complex systems."[23] He acknowledged that much of what we learned was by responding to what the system was showing us. Having more specialist scientists inside, for shorter periods, as was decided for future closures starting with the second, would enhance our ability to take advantage of these opportunities.

Cooperation Through It All

Roy came to laud our crew—giving credit to its heterogeneity and adaptability through our rigorous, if unorthodox, training program. The fact that our team was four women and four men and from widely different cultural and personal backgrounds helped. We operated to a large extent as a "work democracy"—everyone had a primary area of responsibility (and often more than one) for which they were the operational and strategic manager.[24] We called for help as needed for our area of responsibility. Morning meetings reviewed operations and made plans for the day, week, and month.

It is striking that despite periods of group stress and division, especially during the period of the outside power struggle, we continued to work together and openly discussed these conflicts. There was never an instance of subconscious sabotage of the overall Biosphere 2 or of anyone else's research or operational areas. Concern for our "lifeboat" and goals overrode all else.[25]

And so, we worked together through it all.

Oleg Gazenko was an SAC member and saw us during our last year and at re-entry. He concluded, "Your difficulties in Biosphere 2 were nothing compared to our cosmonauts."[26]

There is also a tendency in isolated, confined environment groups to exaggerate the seriousness of group dynamics and psychological issues.[27] Jane titled one of her chapters "Starving, Suffocating and Slowly Going Mad." Although a couple of the biospherians reported feeling depressed and received private counseling through phone sessions, when the crew was given psychological tests, there was no evidence of depression.[28] In fact, we scored very similarly but higher than astronauts, and the profiles of the men and women of our team were quite similar. As a group, we fit the profile of "adventurer/explorer."[29]

The head of the psychiatric department at the University of Arizona College of Medicine, Dr. Alan Gelenberg, held private phone sessions with each of us. Rumors were flying that we were going nuts. When our outside shrink finished, he told us we were in good mental health. Dr. Robert Bechtel and Dr. Michael Berran of the University of Arizona had us complete the MMPI (Minnesota Multiphasic Personality Index) while inside. They concluded that we all showed a high degree of independent-mindedness and resourcefulness. Berran also commented, "If I was lost in the Amazon and was looking for a guide to get out, and to survive with, then you'd be top choices!"[30]

Ups and Downs and Special Occasion Truces

We benefited from having a beautiful world and a diverse, healthy diet. We laughed at media reports of a NASA CELSS project that was crafting a diet with just three food crops. Even with dozens of crops, we craved new tastes and food surprises. It was striking that no matter what the group dynamics were, there always seemed to be a suspension of conflict whenever there was a special event, especially one involving food and drink!

Yes, it could be counted on—feast days, birthday parties, coffee mornings—we would all have a great time, hanging out and swapping stories and anecdotes from our shared life and prior experiences. People have an innate desire to be happy; it's the human default setting. So even though our disagreements and personal conflicts remained and perhaps deepened, everyone wanted to have a good time and take advantage of special occasions and limited home-brewed alcohol to have a party, dance, and make or listen to music.[31]

Some studies show the third quarter of an expedition is the toughest. Ours was from September 1992 to March 1993. Our crew was divided on whether it was true in our case. Some said yes (me included), and some thought the first quarter when we were adapting to becoming biospherians was the most difficult. There is definitely an adaptation factor; it takes time, depending on how different the environmental conditions to which you're adjusting, to settle in. We also had a harder time during winter months, both because of the well-known seasonal adaptation disorder (SAD) and because CO_2 rose and food crops and biomes grew slower in the shorter day lengths.

It was also during our third quarter that the battle peaked with our SAC over project management. The consequent polarization of the crew was at its most intense. That was also when oxygen decline was greatest, adding to the limited energy available while eating a calorie-restricted diet. The power struggle finally ended with the project's board dissolving the SAC. The majority of its members chose to continue working on the scientific program at Biosphere 2 by the beginning of our final six months inside.

Oleg Gazenko's simple test for whether a team has adapted to a new environment was a feeling of freedom. For me and most of the crew, that was definitely the case.

Near Mission-Ending Accidents

There were several accidents and technical failures that could have abruptly ended the two-year closure before completion. It was never a sure thing that we'd stay inside for two years. Laser, our emergency captain, was great at surprising us with medical and fire emergency drills when we got too complacent. It wouldn't take much to end the mission.

The first winter, a fire nearly engulfed our carbon dioxide scrubber when a three-prong electric cord was incorrectly wired and left atop a tank. When Taber happened to pass by, it was just starting to melt through the plastic. Flames were beginning to flare out of the smoke. Had it caught on fire, it would have catastrophically polluted our small atmosphere.

Another time, Roy left something on the hot plate of the medical lab and locked himself out. Unable to get back in after getting instructions about finding a spare key, he dozed off until morning, leaving the hot plate on. Fortunately, no fire was started. Linda almost went into anaphylactic shock after an allergic reaction, but Roy had an antidote in the medical lab. There were falls and bruised backs; I feared a concussion when I slipped on a concrete floor. No one fell off the savanna cliff, rainforest mountain, or from a space frame, though that would have resulted in serious injury.

It's time for me to finally confess I nearly cut my finger off. I was hell-bent on my campaign with the victory gardens. I was working overtime one Saturday evening, cutting pieces for new planters on our table saw. I had run a woodworking enterprise for years in the early 1970s at Synergia Ranch and prided myself on my almost morbid care when using power equipment. But for one mindless moment, I was careless and could feel part of my thumb taken by the blade. Visions of Jane's finger and its worldwide media sensation a year earlier flashed through my mind.

Oh my God, I'd thought. *What have I been stupid enough to do?* When I gathered my courage, I finally dared take a look. . . . It had just grazed one edge of my finger nail flesh. I wrapped up my badly bleeding finger and told no one. Days later, when healing was underway, I let Roy give me some antibiotics, though I never told him how I'd injured myself.

The closest call came three months from the end. Around 5:00 p.m. one afternoon in early June 1993, a series of unlikely events set the stage. A rare bushfire cut highway traffic and damaged the power lines to Biosphere 2. The power cut should have automatically activated our generators. One of our three generators was down for maintenance; another couldn't be started for unknown reasons. When the team at the energy center got the remaining generator running, the main breakers kept shutting down. There was no electricity to Biosphere 2.

Inside, it was all hands on deck. Linda was monitoring and calling in temperatures from the rainforest, our most sensitive biome. Outside, they tried bringing the power to the Biosphere back online one sector at a time. Crew members ran around closing down air handlers and pumps to lessen the load. The circuit breaker refused to reset. Linda reported that we had climbed from 99 to 104 degrees. We started to review evacuation protocols.

Bill Dempster, who was the overall "Mr. Fix-It" outside, just like Laser was inside, was finally reached by cell phone in town. He tried to walk the energy center team through a reset. After hearing the circuit breaker clunk on and off a dozen times, he told them to stop. Repeating what's not working does not eventually lead to another result. He advised them to recheck the operating manual. Sure enough, there was something else that had to be done before the circuit breaker would reset.

In moments, power was returned and temperatures began to drop inside.

But, if this had happened at midday or early afternoon, the more than an hour it took to regain power could have been fatal. Our inside temperatures might have rapidly hit 150 degrees, badly damaging plants. To avoid heat-stroke and worse, we would have had no choice but to leave.

Caring for Our Lifeboat: We're All in This Together!

We loved Biosphere 2; we had seen it grow from an idea to a construction project, to our home and world. In a paper co-authored by four of the bio-spherians about the human factors in exploration, it was noted: "All the crew knew that anything which hurt living or technical systems might quickly and directly imperil their own health. [We] kept overall Biosphere 2 air and water quality, carbon dioxide and oxygen levels in constant attention, in a very visceral and profound way, not just as a mental abstraction. This intimate "metabolic connection" enabled the crew to discern and respond to even subtle changes in the living systems."[32] Similarly in a review paper on the lessons from Biosphere 2, which I co-authored with Kathelin Gray and John Allen, we emphasized: "Appreciation of the value of biosphere interconnectedness and interdependency was appreciated as both an everyday beauty and a challenging reality."[33]

This cooperation overrode personal animosity and discord with project management. Roy acknowledged, "I don't like some of them, but we were a hell of a team. That was the nature of the factionalism . . . but despite that, we ran the damn thing and we cooperated totally."[34]

Solastalgia and the Fight to Preserve Our World

There are parallels between our Biosphere 2 experience and ways of realigning with our global biosphere. Even though there may be conflict and disagreements, like the divisions that beset our crew, it does not preclude a deeper connection to and responsibility to the biosphere, to nature. This manifests as the love and bonding that so many people and cultures feel with their part of the world.

In ecopsychology, *solastalgia* is the distress felt seeing environmental damage to a place one loves.[35] The term resonates with so many around the world. We are in grief and turmoil over the state of the world and our local environment, even if we're not fully conscious of it.

All over the world, people are fighting for their communities, their lands, their part of Mother Nature. They are transcending "Not in my backyard" to "Not in anyone's backyard," making alliances between cultures and continents.[36] Though the forces of exploitation may appear invincible, powerful countervailing forces are deeply rooted in our humanity and evolutionary history. These include *biophilia*—our innate love and connection to life and living systems[37] and *topophilia*—our love of place and landscape.[38]

As Naomi Klein puts it in her survey of the current ecological crisis, "the solution to global warming is not to fix the world [e.g. geo-engineering], it is to fix ourselves."[39] We have to change a dominant mindset from "extractivism," which operates as if nature and our biosphere is an infinite resource we can exploit without limit. Instead of the illusion of the "control of nature," a reverence for and a more gentle participation with the world must be rekindled. These attitudes run deep in the now-dominant global culture driven by the imperatives of our economic system for profit maximization and endless growth of consumption.

The Ancient and Modern Wisdoms

The eight of us in Biosphere 2 sometimes felt we were the natives of our world as its first human inhabitants. We were a new tribe with the amazing experience of growing up alongside our living companions—all the plants and animals of the biomes and farm, the microbes in our soils and waters. We knew they had an equal right to life. In fact, it was only their life and vitality that made our lives possible and healthy.

There is deep wisdom in indigenous cultures about the interdependence of all life. Almost universal is the appreciation of their homeland and the Earth as sacred, imbued with the divine. This relationship is quite different than thinking the Earth was created for us.

Contrast that with the more humble (and reality-based) attitude of indigenous peoples. The Nez Perce tribe says: "Every animal knows more than you do." Sioux Indians affirm that "with all things and in all things, we are relatives." They lived in an abundant world, only taking what they needed. "The frog does not drink up the pond in which he lives."[40]

The fight for a more sustainable, beautiful, and peaceful world is about regaining and acting on that sense of the sacred. We are all aboriginals of our home planet. Albert Schweitzer thought all human morality was rooted in Reverence for Life, that "good consists in maintaining, assisting and enhancing life, and to destroy, to harm or to hinder life is evil."[41]

The love all humans have for home and homeland—and by extension, for the Earth—can serve as the guideposts for a path to a better future and cure for solastalgia. The cure to that near universal modern illness lies in fixing ourselves by rebuilding connections to the natural world.

Jane put it succinctly: "We do not need to become stewards of nature; we need to become stewards of ourselves."[42] The eight of us did become stewards of ourselves. The one unthinkable crime that no one was ever tempted to do was to harm our living Biosphere 2 world. On the outside this will mean letting love for the Earth motivate, direct, and inform our actions.

16

Connected!

The Biospherian Experience

Test Module Transformations

IT WAS TWENTY-FOUR HOURS in the Biosphere 2 Test Module that convinced me to become a biospherian.

We had three test module human closure experiments in 1988 and 1989. First John Allen went in for three days, then Gaie for five, and finally Linda for twenty-one days. As a kind of joke, the experiments were named Vertebrate X, Vertebrate Y, and Vertebrate Z. Their closures followed more than a year of ecological testing.[1] Our first constructed wetland, covering all of fifteen square feet to handle the wastewater of one occupant, and a soil bio-filtration planting bed were tested there.[2]

We had no idea how humans would handle being in a closed ecological system, nor its dangers. For safety, Allen wore a finger alligator clip with sensors showing pulse and dissolved blood oxygen while asleep. Dr. Dan Levinson, from the University of Arizona College of Medicine, slept onsite. The environment was also heavily monitored. Allen's response was unexpected.

Already a strange partnership has started building between my body and the plants. I find my fingers stroking, feeling the soft rubbery texture of the spider plant, knowing it's picking up outgassing products . . . Notice my attention

Figure 62. Vertebrate X, John Allen, during his three-day closure in the Biosphere 2 Test Module. This was our first experiment with a person in a closed ecological system.

turning more and more to the condition of the plants . . . I've always had the sense of plants being alive, responsive, even a living symbol. But now they're necessary . . . and since they're necessary, I look out for them. . . . It appears we're getting close to equilibrium, the plants, soil, water, sun, night and me. . . . in communication with the world by sight and sound, but touch, taste and smell all different.

John said when he exited the test module: "I knew with my body as well as with my intellect and emotions that Darwin and Vernadsky were right about the power of the force of life."[3]

After the human closures, attention shifted to finishing Biosphere 2. The project offered staff members twenty-four hours inside the test module. I signed up. Hell yes, I wanted to have a personal experience of what this was all about!

It's All in the Body

The smell and sensations of being in a new world hit as soon as the air-lock door closed behind me. Pungent aromas surrounded me, as the small volume was packed with plants, all kinds of plants. We had minibiomes, so there were rainforest, savanna, and desert species. The thicker air made the experience of breathing different, and my body sensed it was taking in something more nourishing than normal. What happened over the next hours was remarkable. The knowledge of my metabolic connection with that world passed from my head to my body and became palpable, sensory. My body got it as I walked around, looking at the plants, breathing in from our shared atmosphere. These plants and I were breathing together. They were collectively my third lung. My breathing and metabolism were helping them, and without them, how would I survive? That cellular awareness persisted through my time inside. Even when I was in the small "apartment" of the test module, I could feel the presence of the other life I had joined. It was a wonderful feeling: my body and I were totally connected with them. It was like joining a symphony of life already in progress, and I had my role to play as well. My presence here had changed this world, but I fit in fine—there was room for me.

Premonitions of things to come: after I went to the bathroom, I went out to check on the plants in the constructed wetland system. I was smiling, and they looked happy too. I was glad to be of service! What a nice feeling—my wastes were their food.

Friends who stopped by could see what I was feeling. I was far more relaxed than I'd been before coming in, positively jovial. Habitual stress and

Figure 63. Dr. Yevgeny Shepelev peering into the life-packed Biosphere 2 Test Module in 1989. How different from his time in a sealed chamber with just chlorella algae as a companion!

grouchiness were dissipated in that sense of belonging. My body was at ease. It was beautiful and clean of pollutants inside. I knew the air, water, and food were healthy; all the life around me was taking care of it. Like on a ship or airplane, there were also the mechanical sounds, fans, motors, electric lights whirring and humming, part of the soundscape of this techno-living world. I sensed I was part of a force much greater; this assemblage of organisms, air, water, and soil was both ancient and eternal and vibrantly alive moment by moment. Everything was in communication, in metabolic interconnection. This world now included and supported my bodily existence.

After I exited the test module, I joined the list of candidate biospherians and started training. Unexpectedly, I was lucky enough to enter with the first team in 1991.

Entering the World of Biosphere 2

The sense of being connected with life does not hit one quite as rapidly and overpoweringly in Biosphere 2 probably because of its far larger size. But it grows over time until you have that bodily understanding. This biosphere, this world of life, is shaping and sustaining my organism, and my body is fully a part of its overall metabolism.

That feeling of connectedness is deeper than just that marvelous cellular sense of metabolic union. It was strengthened from the roles we humans played in Biosphere 2.

Biospherian responsibilities encompassed jobs usually done by separate categories of people. We were farmers, laborers, technicians, researchers, and managers, often all in the course of a single day.

We were the farmers of our world. We planted, tended, and harvested our crops. That farming allowed us to eat and was an important part of our world's metabolism, affecting air and water quality. Knowing our farm had been designed to be fully integrated with the rest of Biosphere 2, that it would produce nothing damaging to us or the rest of its life, made being farmers hard labor but incredibly satisfying. We biospherians rarely forgot. We knew the full history of every item in each dish of our meals. We assisted in getting every crop to the dining room table.

We were the technicians and engineers of our world. Checking, calibrating, maintaining, and sometimes repairing sensors and equipment was up to us. We had a well-equipped workshop where Laser, the techno-magician, could build needed equipment or perform repairs. Many key parts of our technosphere were displayed in real time with sensors on computer screens. We were ever aware of all that nature ordinarily does and what we had to replace with technology. Perhaps that increased our sense of being stewards, assistants to all the life inside. Like us, we'd think, you've been cut off from Earth's biosphere, which provides weather and seasons, winds and waves, and the relative safety that comes from being part of a so much larger world. But you can count on us to work on your behalf. One of our most important jobs was to make sure that all the nature-replacing technology works, making possible the life of Biosphere 2.

Our roles as researchers and safe-guarders of our own health and the biosphere's health overlapped. We diligently monitored what was in our air and water. Other research projects included tracking ecological developments and changes in our biomes, curious, like the rest of the world, what would happen. We and our biomes were like children growing up and learning how to live together in a new world. Would Biosphere 2's biomes become overrun with urban weeds like Lynn Margulis predicted?[4] Would we lose 80 percent of our species, like H. T. Odum thought? We took our jobs as keystone predators very seriously. We worked—many hours every week, as it turned out—protecting biodiversity and helping our biomes grow up.

The Wild Out There and in Us

Wilderness and the unstoppable force of life are not just "out there," but they are in us as well. We are part of the world of life. Conversely, the degradation of once pristine areas is felt in the human heart and spirit, even if unconsciously. Having the wilderness areas in Biosphere 2 made all the difference in the quality of life we enjoyed. They were our Yosemite, our Grand Canyon, and our Central Park as well.

There was this strong psychological sense of being inside when we were in the human habitat, and a sense of being outside when we were in the farm or the biomes. Like any visitor to a national park, being amid that greenery and beauty nourished our spirits. Watching the day-to-day and seasonal changes in those biomes we loved and protected was a never-ending delight. We were proud, like parents are of their children, when we prepared to leave. We saw how our biomes had grown up, just as we had, during our two-year journey together.

The smallness and fragility of Biosphere 2's minibiomes meant that there was no alternative to some degree of human management. We joked at morning meetings that we had "weather requests" more than weather reports. Our conversations went like this:

Linda plans to water in the lower rainforest today at 3 p.m. for forty-five minutes. Does anyone have a problem with that—does the schedule need to change? If not, if you want a walk in a rainy rainforest, that's your time. Laser will let mission control know that nighttime temperatures in the savanna could be brought down 2 to 3 degrees, and to make sure we keep the humidity below 40 percent in the farm.

We all had a strong sense that as important as our roles were, the ultimate force keeping our world—and us—healthy was beyond our control. It was nature, just as it is in our global biosphere. There was strong consciousness of our debt to our fellow life. Just as John Allen reported in that first test module closure, it was instinctive that we took care of our plants and were never careless in how we walked through them. In half an hour, we could take a trip around our world and thank all the plants for what they did, the soil microbes for their contributions, for keeping us healthy and for their beauty. We knew we were surrounded and protected by both the power and the intelligence of life.

Although Biosphere 2 was remarkably free of pollution, analysis of our blood found something quite surprising. All kinds of environmental pollutants, including ones that had been banned long ago like DDT, were spiking. Losing weight and body fat released contaminants stored when we were younger. These levels decreased later in the closure.[5]

Bonding with a Biosphere

Our two-year voyage in a time-machine radically transformed much of who and what we were. I think of it as the Biosphere 2 purge or cleansing. Gone were assumptions of plentiful available food, of oxygen-enriched air, of being the beneficiaries of our biosphere without having to get engaged with it, working to keep it healthy. Gone was the assumption that water simply came out of faucets and pipes, without knowing its quality and where it came from, or where it and our wastes went.

Some lamented there would be no mystery inside Biosphere 2, since it was a human-designed and human-built world. Some suspected megalomania, a project to control nature and make technology supreme. They were far from the mark. In designing the project, we learned that much was unknown about the basics of life. Living in it, we experienced daily the mystery and magic of life.

A desert or thorn scrub plant has a mysterious intelligence—it knows when to swell its buds or drop its leaves. We sharpened our skills as naturalists, with help from outside scientists, reading signs from indicator plants, when to start or stop the rains, responding to the subtle signs of organismic and ecosystem health. The mangroves and corals defied predictions about their adaptation to high elevation in a different part of the world.

How did the life inside organize itself into such distinctly different biomes in such a small space? I'm quite sure that, blindfolded, all eight of us could tell which biome we were in by its distinctive smells and atmosphere. That mix of quantitative and naturalist science helped us do our jobs well as stewards of our world. We could walk—or snorkel—through our biomes and sense their health and problems as well as check our sensors and data.

Through some management oversight, Kevin Kelly, the technology writer, was left to spend a few hours alone inside Biosphere 2 in spring 1991, months before closure. He went around what Tony Burgess had poetically called our "cathedral to Gaia" experiencing and pondering. He wrote later:

> Great things will be learned inside Bio2 about our Earth, ourselves and the uncountable other species we depend on . . . Already it has taught me, an outsider, that *to live as human beings means to live with other life*. The nauseating fear that machine technology will replace all living species has subsided in my mind. We'll keep other species, I believe, because as Bio2 helps prove, life is a technology. Life is the ultimate technology. . . . because of its autonomy—it goes by itself, and more importantly, it learns by itself.[6]

Theories abounded about the "ninth biospherian." Were the galagos the ninth biospherian? Roy thought the biosphere itself was the ninth biospherian. I personally liked another viewpoint. The eight of us collectively lost 160 pounds of body weight, equivalent to another crew member. Though those molecules were no longer part of our bodies, they were part of Biosphere 2. Linda observed: "The Biosphere owns those molecules."

Those carbon, nitrogen, oxygen, and other elements were cycling around our world, incorporated in soils and waters, corals, concrete, sweet potato vines and trees, in the air at one moment and somewhere else the next. They would do that for as long as Biosphere 2 remained a closed system. We would be replaced by another crew, but our ghost atoms, the ninth biospherian, would remain in the metabolic dance, connected to this entire world. How astonishing—and true—is that?

This is also true for us in our global biosphere. Our atoms were forged in supernovae explosions of ancient stars, and our bodies are in intimate and endless metabolic cycles with everything on our tightly sealed Earth. After our deaths, every molecule that was once "us" will remain in the cycling biosphere dance.

Our connectedness, of course, extended to our crewmates. We were so dependent on one another, to the skills and dedication we all brought to the task and to our shared life. Perhaps the immensity of the task helped hold us together through good times and bad. We were able to successfully operate Biosphere 2 because our crew formed a synergy with project managers and our network of consultants far greater than what we could have done as eight individuals. We all had a strong bond with Biosphere 2 and, ultimately, with one another. We knew everyone was doing their job, lending a hand and a mind when called for, doing what needed to be done and often going beyond the call of duty.

I wrote soon after leaving Biosphere 2: "When you live in a world you can traverse in a leisurely stroll of 15 minutes, this mutual dependence becomes a fact of your life and changes how you think and act . . . Do good, and the biosphere thrives. Be thoughtless or foolish, and your bad action will adversely impact your own life. Every action counts, and with only eight people in your world, acting intelligently and cooperatively is not a luxury!"[7]

No action was anonymous in Biosphere 2; there were no actions too small to count. Everything we did had an impact. This leads to a wonderful ecological mindfulness, not as a restriction of one's freedom, but as recognition of our interconnectedness and responsibility for our world.

New Roles for Humans

I've met people who regard humans as a cancer in the global biosphere.[8] But humans are an evolutionary product of the biosphere just like beloved ecological icons: elephants, tigers, pandas, whales, corals, redwoods, saguaro cacti. Humans have an unusual set of skills, highly developed language and symbols, intricate cultures, and technological inventiveness. We may also have an innate urge to "disturb the universe," as Freeman Dyson puts it.[9]

These abilities can be applied to support and enhance life; or for their opposite, amassing personal or national wealth and power at the cost of deterioration of the biosphere. Our technological wizardry means there is no definite carrying capacity for human population or limits to our supplanting of natural biomes. We may find those limits by pushing present trends and expanding until we collapse. Extrapolation says it will happen.

A few months into closure, I had lost about twenty-five pounds and weighed what I did when I was eighteen years old. Extrapolating that weight loss over the two years, I'd come out weighing minus ninety pounds! Of course, that extrapolation was absurd. It didn't anticipate the increased metabolic efficiency our restricted-calorie diet produced, or how hunger increased our farming skill and ingenuity. We're in roughly the same situation with our global challenges. It's almost impossible to extrapolate current trends because they make no sense. As the economist Herbert Stein noted: "Things that can't go on forever, don't."[10]

A more positive future will unfold if we humans are indeed on the verge of a historic change of perspective. There are signs of a widespread shift of attitude toward our biosphere. The bonding and connectedness we biospherians experienced were based viscerally, in our bodies, on our dependence on our biosphere. We also knew we were neither a cancer nor parasite. We played key roles in keeping our biosphere and ourselves healthy.

Gaie recalled:

We all, all eight definitely said it, time and time again, that we realized the health of the biosphere was our health. If our biosphere was healthy, we were. And that's something that doesn't happen out here [in Biosphere 1]. We don't live that day to day; we appreciate that knowledge, but we don't live it.[11]

Marshall McLuhan said, "There are no passengers on spaceship Earth, we are all crew." How satisfying to be part of that crew in our minibiosphere and to learn to do our jobs well. Atmospheric managers, keystone predators on the side of biodiversity, operators of a technosphere in the service of life, healthy farmers who knew keeping our food and water healthy was a necessity, not an indulgence. These are roles that can inspire all people. These are stories with fulfilled people and life-reverent evolution as outcomes. Bonding and mindfulness are what we need to change how we humans interact with Earth's biosphere.

Celebrating Our World

I should emphasize another role we, as biospherians, played. We were the observers and celebrators of our world. We wrote poems, made documentary films, composed and played music, painted, danced, and sang in celebration of the beauty, wonder, and gifts of our world. We held perhaps the first two interbiospheric arts festivals through video links to share with outside artists, poets, and musicians. Roy collaborated with performance artist Barbara Smith in a number of electronic link-ups while she circled the world.

We see this in the wisdom and folktales of cultures around the world. They celebrate and pay reverence to the natural world and the ways of life of people living in harmony with its bounty. The novelist William Burroughs noted, "the way to kill a person or a nation is to cut off their dreams, their magic."[12]

We had a dream. World leaders, instead of meeting in posh conference centers and staying at expensive five-star hotels, would spend some time in a minibiosphere before making decisions that affect the world's people and the global biosphere. It's hard to imagine they would then think protecting the environment is a luxury that comes after every other priority, or only when they felt affluent enough. A few days inside a minibiosphere might significantly change their worldview!

Gaie and I wrote in a paper for the Space Studies Institute just before our two-year biosphere experiment ended:

> The vision of Gerard O'Neill and others who have contemplated human destiny in space foresaw the evolution of *Homo cosmicus*—humanity no longer confined to one planetary surface. We can add from our foretaste in Biosphere 2 that *Homo ecologicus* will also be an evolutionary product . . . humanity in space will know from the beginning what it has taken us so long to learn on planet Earth—that our own fate and that of our biosphere are inseparable.[13]

17

Re-entry

Anticipation of a World-Changing Moment

OUTSIDE, A CROWD of thousands had assembled. An orchestra was playing Vivaldi's *Four Seasons*. The event was being beamed around the world to as many as a billion people. Luminaries were speaking on the podium. Inside, we lined up in reverse order from how we had come through the airlock two years earlier. We listened to the proceedings going on outside via our two-way radios and prepared for an out-of-this-world and into-that-world experience. September 26, 1993, had finally arrived.

I certainly had mixed emotions. As eager as I was to see loved ones and experience life outside our confined world, I was sad at the prospect of no longer living in Biosphere 2. In the biosphere, everything made sense and had significance. My connection to and responsibilities in working for the benefit of our living systems were clear and immediate—and profoundly satisfying. I had even quietly informed project directors that I would be willing to join the second closure team if necessary. A second closure was scheduled for March 1994 after five months of transition research, improvement of the system, and training of the new team was complete.

We were all excited anticipating feeling the differences between the two biospheres. As soon as we exited through the airlock and took our first

Figure 64. Some of the people gathered to give talks and help us celebrate re-entry. From right: John Allen, marine biologist Sylvia Earle, IE director Robert Hahn, chimpanzee researcher Jane Goodall, Harold Morowitz from George Mason University, and Dr. Leonid Zhernya and Oleg Gazenko from IBMP, Moscow (photo by D. P. Snyder).

breaths of air, I could sense the change. The outside air was thinner, less "tasty" than Biosphere 2's, and filled with different aromas. From a moist pungent tropical air reflecting the intensity of life inside, we now breathed a drier, high-elevation desert air. No wonder, I thought as I looked up at the vastness of the blue sky overhead, not limited by space frames and glass.

It was a very emotional morning. We had all prepared some short comments, but I had allowed for some improvising as I wanted to incorporate the moment as it unfolded. As I stepped to the podium when it was my turn and looked out at the gathered people, I was momentarily overcome. The quality of light as you looked toward the horizon had subtle spectral gradients I hadn't seen for two years. It was just a second or two. I've watched the video of re-entry and can see that palpable moment of sheer delight, looking out and taking in the beauty of our world. In Biosphere 2, we could, at most, see a few hundred feet away. It was good to be back in Earth's biosphere, our home.

It was also amazing to not have a barrier between us and other people. I could feel the emotions and presence of those thousands of people as powerful pulsating waves, enveloping us. It was a delicious experience feeling and sharing their excitement and support.

Re-entry Talks

John Allen at his re-entry talk spoke about the ethics (which he defined as doing what you ought to do, not necessarily what we would impulsively like to do) developed in Biosphere 2.

> The eight biospherians ate, slept, worked, dreamed, enjoyed and suffered, in short existed in harmony with their biosphere. Their biosphere flourished with their way of life, they recycled their food, their wastes, their water, their air. They protected biodiversity and enhanced the beauty of their landscapes. Their own bodies purified and their biosphere sparkles undimmed without a ghost-like fog of smog. They lived with high tech instrumentation and communications but in a non-destructive, ecotechnic way . . . I appreciate the biospherians' skill in operations, their integrity in research, their zest for exploration but I honor them for their ethical achievement, achieved at no small cost to their immediate gratifications, for having done what they perceived they ought to do.

Listening again to the short talks of the biospherians, I'm struck by how many talked about the striking differences between Biosphere 2 and the outside world and the profound personal change they'd gone through in being so connected to a living world.

Gaie, the first of the eight of us to speak, was obviously somewhat overcome at first. She started by saying: "It's a different atmosphere; it's really a very different atmosphere! For the first time in two years, I've looked back on Biosphere 2. For two years I've been living in a different world, a different atmosphere, a very different existence . . . Looking back on it, it's a symbol of great beauty, of the merging of ecology and technology. Ecology has been studied without the inclusion of man and man's technology. That has now been fundamentally changed. There's an important new perspective."

Roy followed: "Any substantial amount of time spent inside Biosphere 2 brings you back to the idea of a natural paradise that Earth could be and should be and we hope will be. One of the biospherian jobs in the future may be not only to build new biospheres but to rebuild Biosphere 1, Earth, our own home planet. We'll need a lot of help to bring that about."

Linda: "This has been an extraordinary experience and a magnificent journey for the past two years. I've also glimpsed paradise in here, glimpsed paradise. . . ."

I said: "They said it couldn't be done, but here we are, healthy and happy . . . operating Biosphere 2 has changed the way I operate my organism. To live in a small world and be conscious of its controls, its beauty, its fragility, its bounty, and its limits changes who you are."

Jane was next: "I, too, have changed in the past two years . . . I have been in charge of agriculture and have thought a lot about food . . . thinking about the first piece of food I'll have. Perhaps bread, and I'll think where did the wheat come from? What pesticides? Where were the chickens kept that laid the eggs in this bread? I'm not going to know the answer to any of those questions. Whereas in the biosphere I certainly knew the answers to all of those questions."

Taber said: "It's been a wonderful journey and we're all profoundly different for having been through it . . . Science, technology, and understanding advance in increments, and I think we're looking at one of these increments. The historic significance of this will not be known for many years."

Sally continued: "I really can't believe it's over so quickly, our two years inside Biosphere 2. . . . What's going to be really difficult is portraying what it was like to be inside. What was it really like to recycle our air, our water, our nutrients. This is the beginning of a lifelong journey of learning how to manage a small biosphere."

Laser was the last to speak:

Wow! Two years ago I stood here wondering what was going to happen. Today I stand here wondering what happened! One thing I know happened is that the water we started with in September 1991 is still the water inside Biosphere 2. It recycles through the whole system. It is not polluted. That water is sometimes drinking water, sometimes ocean water, sometimes wastewater, subsoil water. It recycles, it's not polluted. Our atmosphere in there is not polluted. We grew our food in there without polluting our own system. For me personally, what I discovered was we really have it in our own hands to either destroy our planet or make it a beautiful system, a platform to step off to explore the galaxy.

Welcome Back

The speeches finished, we climbed aboard a golf cart to be driven to mission control for debriefing. Though the project had a few solar-powered golf carts, someone hadn't thought this one through. This one had a gasoline engine. Our broad smiles at the sensation of speed (riding a horseless carriage!) turned to disgust, inhaling those noxious exhaust fumes like everyone else!

It reminded me of the scene when Akiro Toyoda, of the Toyota car company his family founded, visited Biosphere 2 with his corporate entourage. He told us he loved the project, with one exception. He said he wished one of his excellent cars was inside with us! We smiled and continued conversing. No one was game to tell Mr. Toyoda that if we ran his car inside, we and probably every other living organism inside Biosphere 2 would become very sick.

Next up was a buffet spread at mission control, a dazzling collection of food, which was strange in a way because we hadn't grown it and didn't know where it came from. It included cheeses, fruits, cold cuts, even bagels and lox: a mountain of food. Afterward, the doctors did complete medical examinations, including measuring our body fat. They were concerned we not be overstimulated our first day back, unaccustomed as we were to mingling with other people and eating exotic foods. With so many unknowns, they played it safe, and had us visit the makeshift medical examining room three times during the day. One of the concerns was us overeating, since our bodies had gotten so efficient at extracting all the nutrients from what we ate. Another worry was exposure to new germs, since it's well known that people in small, isolated groups wind up with reduced microbial diversity. As a consequence, they don't have immunity that people in contact with greater numbers develop and can be susceptible to germs they haven't been exposed to.

A press conference followed, and then crew photos, with much jostling and elbowing by photographers, which added to our amusement. From my journal:

> The press conference was funny—I felt euphoric and giddy, not in any mood for usual press questions—and with a reawakened sense of humor. When

asked whether the contact with so many people was disorienting or intimi-
dating, I replied that the emotional connection was electrifying and somehow
far greater than with a glass barrier. That I was feeling it even in a tent filled
with journalists who despite that profession remained human (that drew a
large laugh). The most curious question was whether the biosphere was trying
to kill us by the lowering of oxygen. I replied that from my perspective it was
trying to teach us something.

The rest of the day alternated between the public and private. It was
amazing to be with so many friends and family again, to be with my eighty-
three-year old Jewish mother, who had escaped the Holocaust by coming
to the New World. After two years of separation, it was special to hold my
then-six-year-old son, Mars, again. We had exchanged emails and occasionally a phone call because international calls to Australia in those days, before
Skype, were quite expensive.

Sally Silverstone and I led a tour of VIPs through Biosphere 2. It was the
first of many experiences during the transition period when I could again

Figure 65. Reunited with my son, Mars Nelson Tredwell, then six years old, on re-
entry day (photo by Phil Dustan).

appreciate how fully adapted I was to the Biosphere 2 environment. When we finished, someone commented how everyone except Sally and I were sweating profusely. For us, it was still more comfortable than being outside. It would be months before my center of gravity slowly shifted, and I felt fully at home outside Biosphere 2, on my learning curve in rejoining our global biosphere. I also noted that neither the visitors nor I had any complaints about the air inside: it was moist, fresh, and filled with tropical life.

Re-adaptation: Now What?

Some things were jarring amid the pleasures of coming home. I took my son to Canyon de Chelly and the Grand Canyon, where we hiked down and overnighted at the bottom. I had a road trip to California with my son and his mother, Robyn Tredwell (a fellow Director of the Institute of Ecotechnics who managed Birdwood Downs in Australia). We went to San Diego and enjoyed the zoo and safari park. But there were problems lurking behind the beauty of the city. When we went to a seafood restaurant, I innocently asked if the fish was local. Our waiter answered frankly: "It isn't local, and you'd better be glad it isn't!" And then there was Los Angeles and driving on freeways with seven or eight lanes in each direction. I had just been getting used to cars again. Were these cities built for the convenience of people or for our cars to enjoy?

My first trips to supermarkets were strangely disorienting. I had taken visiting Soviet scientists to them before closure. For most, it was their first visit to the airplane-hangar vastness of an American food store. They were overcome with the immensity of choices and types of foods on the shelves. One almost sobbed as he quietly lamented, "My poor, poor country," thinking of the paltry grocery stores back home.

I was shocked seeing so little fresh food and so many packages—cans, boxes, cartons. But it was nice that all I needed was some cash or a credit card and they gave me food. I didn't have to change into work clothes, grow it, or process it! So many nearly identical versions of the same thing. How do you decide? I read with horror the lists of ingredients on the boxes' sides—a chemist's dream list—then took the shopping bags home, unloaded the goodies into the fridge and pantry, and contemplated the enormous pile of packaging. Two years without packaging or trash.

All the while, I continued to go into Biosphere 2 daily, helping with all kinds of research projects underway and helping train the next crew and pondering how I could be helpful in our global biosphere. How could I remain aware that everything we do has consequences, how could I remain connected even though this biosphere is so vast?

18

The Afterglow

Returning to the World

SIX MONTHS INTO CLOSURE, a group from the Mexican Association of Journalists asked what I would miss most when I leave. I answered: "I think I'll miss the consciousness of being responsible for my actions and their impact on my surrounding life systems. But I'll try to keep that state of mind when I go back into Biosphere 1."

The crew watched two PBS series on DVD with mythology scholar Joseph Campbell. We were struck by the "hero's journey." We didn't see ourselves as heroes, but the basic mythic theme resonated. The journey starts with a call to adventure. On the journey, suffering and personal weaknesses must be overcome in a magical, otherworldly realm. If successful, this leads to spiritual death and rebirth, personal transformation. The return journey is hardest, re-entering the ordinary world. There, one learns how to share insights and work for the benefit of everyone.

We left Biosphere 2 and had to figure out how to best use what we had learned there, who we had become there, and how to pay back the world.

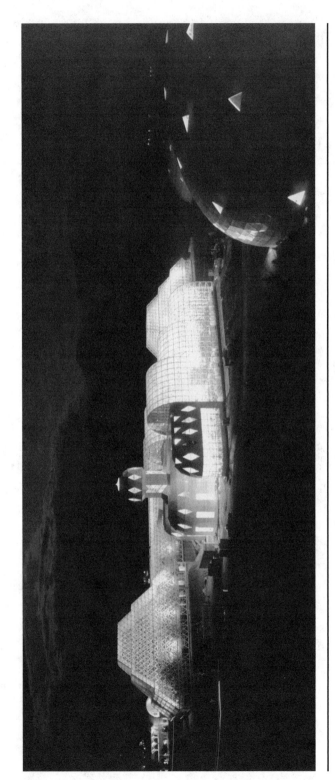

Figure 66. Biosphere 2 lit up at night (photo by Gill C. Kenny).

Becoming a "Wastewater Gardener"

I decided after Biosphere 2 to continue my work with the Institute of Eco-technics. This included organizing conferences and helping manage the fruit orchard at Synergia Ranch, the pasture regeneration at Birdwood Downs in Australia, and sustainable rainforest research at Las Casas de la Selva in Puerto Rico.

But I also wanted to deepen my knowledge and continue developing constructed wetlands for sewage treatment. This technology excellently illustrates the kind of paradigm shift we need. Biosphere 2 visitors helped me realize how new this approach was. At first, others were incredulous then later delighted when they finally understood that the flowering verdant constructed wetlands (and the smiling person tending them) were our sewage treatment system.

For me, a crucial additional benefit of this kind of approach is that it connects people to a part of their ecological reality. Since it is done on a local level—from individual house to community—it removes the mystery

WASTEWATER GARDEN SYSTEM®

NB: depending on project, graywater can be separated via a sedimentation tank and/or grease trap (in case of heavy kitchen activity - restaurants for example) and fed directly into drainage / subsoil irrigation / green filter trenches.

Water entering

Control Box: Can be placed inside or outside the WWG

Graywater

Gravel

Blackwater

Filter

Gravel

Soil

PRIMARY TREATMENT
Septic tank or similar
Residence time: at least 2.5 days

SECONDARY TREATMENT
WWG TREATMENT UNIT
Residence time: at least 4 days

FINAL RE-USE DRAIN FIELD
Subsurface irrigation for
additional productive green zone

Figure 67. Schematic showing the steps in a Wastewater Garden: septic tank, constructed wetland, and final subsoil irrigation. M. Nelson, *The Wastewater Gardener: Preserving the Planet One Flush at a Time* (Santa Fe, NM: Synergetic Press, 2014).

Figure 68. Sunrise School WWG, Kuta/Legian, Bali, Indonesia. This system treats the black water of seventy-five students and teachers and serves as an outdoor classroom (photo by Emerald Starr). M. Nelson, *The Wastewater Gardener: Preserving the Planet One Flush at a Time* (Santa Fe, NM: Synergetic Press, 2014).

of where wastewater goes when toilets are flushed. When people realize the beautiful wetland they can visit and enjoy is thriving because of their waste, it is truly satisfying.

This inspired me to return to academia after an absence of twenty-five years. I received my master's degree from the School of Renewable Natural Resources at the University of Arizona. For my thesis, I designed a zero-discharge, constructed wetland using fast-growing poplar and willow trees. Wanting to work with H. T. Odum, I did my PhD research at his Environmental Engineering Sciences Department at the University of Florida.

Going back to university reminded me that conflict, discord, and gossip are found in all groups of people, including faculty! A Columbia University political scientist had quipped, "Academic politics is the most vicious and bitter form of politics, because the stakes are so low."[1]

I was inspired studying beautiful thousand-acre constructed wetlands near Orlando used as recreational parks. There, visitors enjoyed the natural setting, most not realizing the wetlands were treating and reusing sewage.

My dissertation research included building two prototype constructed wetland systems with the Planetary Coral Reef Foundation, a division of the Biosphere Foundation, in Akumal, Mexico. The foundation was started while in Biosphere 2 by my biospherian friends, Gaie, Laser, Sally, and John Allen, with the assistance of Space Biospheres Ventures. Akumal is where we collected most of the corals for Biosphere 2, so it repaid a karmic debt to bring an ecological system back to the region. Our constructed wetlands protect the beautiful reefs from sewage damage. I used a nearby mangrove wetland as a final biofilter for the treated wastewater. We built more than thirty constructed wetlands along the Yucatan coast afterward.

I call my systems "Wastewater Gardens." We innovated by using a wide biodiversity of plants, including ones that can be harvested, as well as others

Figure 69. Wastewater Garden for the Centro Ecologico, Akumal, Mexico, 2014, eighteen years after it was installed (photo by Gonzalo C. Arcila). M. Nelson, *The Wastewater Gardener: Preserving the Planet One Flush at a Time* (Santa Fe, NM: Synergetic Press, 2014).

for their beauty. It's sad that most constructed wetlands are designed by engineers with no appreciation for creating microecosystems to provide wildlife habit and biodiversity. Most constructed wetlands are monoculture "reed beds" with one or two wetland species. Beauty is an important element; it triggers biophilia (love of life and the living world). People love green plants; we have a harder time with anaerobic bacteria in a septic tank. Bonding with the plants in your Wastewater Garden makes you more mindful about not flushing harmful chemicals down your drains.

After graduating, I headed the Wastewater Garden division of the Biosphere Foundation. They promoted the technology for earlier projects, including many in Mexico, Bali, Indonesia, Puerto Rico, and Australia. Later I started a company, Wastewater Gardens International, to continue the work. With regional affiliates, we've implemented more than 150 Wastewater Garden systems at varying scales in thirteen countries around the world.[2]

Life after Biosphere 2

My seven crewmates continued to do outstanding work, much of it inspired by their experiences and transformations in Biosphere 2.

Roy returned to his pioneering nutritional and antiaging studies at the UCLA Medical School. He and colleagues published ten scientific papers on the results of the Biosphere 2 diet, lowered oxygen, and other biomedical studies. He also continued his work in performance art. He died in 2004 of ALS at age eighty. In light of his diagnosis, probably much of the health decline he suffered in the latter part of the closure was due to early impacts of ALS. Posthumously, the American Federation for Aging Research and Integrative Medical Therapeutics for Anti-Aging recognized his work with awards. Roy made a documentary film about Biosphere 2 and acknowledged he had been wrong in his narrower attitudes about science during the closure.[3]

Taber and Jane founded Paragon Space Development Corporation while still inside, enlisting fellow graduates of the International Space University and other space engineers. They got married at Biosphere 2 shortly after closure. Their space company does innovative space and high-stratosphere projects. They studied a laboratory-sized, closed ecosphere on Mir Space Station and a NASA shuttle flight. The systems included small crustaceans, aquatic plants, microbes, algae, and water. That was a milestone—the first closed ecological system in space. Their experiment was also the first to successfully breed aquatic animals in the microgravity of space.

Paragon's stratospheric exploration division developed a self-contained space suit and recovery system. They set a world record with a jump from 135,000 feet attaining a vertical speed of 820 miles per hour. The expertise has allowed them to offer a new experience in near-space tourism—Worldview Enterprises. Paragon has also undertaken many projects related to environmental control systems on space craft, space suits, and capabilities for space habitats. Jane and Taber continue to dream and engineer for space. They're currently candidates in one venture proposing a human spaceflight orbiting Mars and back![4]

Linda received her PhD researching the Biosphere 2 rainforest, working with H. T. Odum at the University of Florida. She teaches ecology, is an environmental consultant, and has an organic worm farm enterprise. As a teacher at Central Arizona College, her science students built "biotubes" in which they lived with plants and soils. Biotubes, made of lumber and plastic sheeting, are dramatic ways to teach all life's interdependence. Linda noted: "It's hard to think of the whole world and wrap your thoughts around how the globe will work. If I put a bio-tube over me, it's a new way of thinking. We're breaking things down small enough to be able to start thinking about them."[5]

In more recent years, Linda and her partner started Vermillion Wormery in Oracle, Arizona, where she lives. Their aim is zero organic waste by using vermi-composting (composting using worms) to eat all kinds of organic "garbage." She also leads workshops and lectures, one of which is titled, "The Green Lessons of Biosphere 2."

Gaie, Laser, and Sally have continued their work with the Biosphere Foundation. Inspired by their work in Biosphere 2, the foundation's mission is to inspire intelligent stewardship of Earth's biosphere. Their projects include community-based and sustainable conservation programs, educational programs aimed at motivating involvement to make a difference, and providing unbiased data about the biosphere.

Marine and coral reef research and conservation continue to be Gaie and Laser's focus. They work at sea and with indigenous communities. After more than a decade of work using IE's RV *Heraclitus* mapping and monitoring coral reefs at more than 140 sites, the Biosphere Foundation now has its own oceangoing ship, *Mir.* The ship's home base is at the Raffles Marina in Singapore. Among their recent initiatives is the Indian Ocean Marine Mammal Research Unit program. They've partnered with universities in Sri Lanka to assess and reduce the number of whale accidents in the shipping lanes. Coral reef conservation continues to be a major part of their work. Some of their recent projects are protecting the reefs of Menjangan Island, Bali, and documenting long-term coral reef decline in the Caribbean and elsewhere.[6]

Sally concentrates her work on sustainable agriculture and ecological protection and conservation. She is the Biosphere Foundation's coordinator for Biosphere Stewardship Education programs. Working in Bali Barat National Park, Sally directs immersive education programs for international and local students there and on neighboring farms. The needs of the local community dictate the choice of conservation activities. Her work with farmers' co-operatives builds on the pioneering agriculture work in Biosphere 2. Sally develops and demonstrates water-conserving organic farming techniques as well as increasing food production by methods to extend the growing season.[7]

All the biospherians have contributed to the legacy of Biosphere 2, publishing scientific papers, books, making documentary films, and giving lectures. I notice no one emphasizes the divisions or interpersonal frictions that seemed so important back then. What we're all inspired to talk about is how we lived with, took care of, and grew to understand how connected we were to the world of life. What we accomplished together with our outside team reduced the rest to a footnote. Roy ends his film with these words: "I predict that Biosphere 2 will eventually come to be recognized as one of the most forward-looking, innovative, and visionary projects of the second half of the twentieth century."[8]

I had the honor of speaking at the Royal Geographical Society in London in 1994 on our expedition living in a new world. We had proudly carried an Explorers Club flag inside during the two years. I looked around at the names of the immortals emblazoned around the hall: Cook, Burton, Shackleton, Scott, Byrd. The most touching talk I've given was in a small, remote,

Figure 70. Fourth International Conference on Closed Ecological Systems and Bio-spherics in 1996 at the Linnean Society, London, where Charles Darwin's first paper on evolution (co-authored with Alfred Wallace) was presented. Nicholai Pechurkin and Lydia Somova pose with author under a famous painting of Darwin.

mostly indigenous town in Chiapas, Mexico, where Hugo Guillen Trujillo, a fellow graduate student at the University of Florida, had grown up. The children came up to me: "You coming here is like an astronaut from the moon came to visit!" I wondered how in such a remote place they'd seen the images of Biosphere 2. This demonstrated again that Biosphere 2's impact was really global, especially for a time before the worldwide spread of the Internet.

When I think of my crewmates, I realize again what an extraordinary team we were. We achieved so much and shared such unique experiences. The Linnean Society of London hosted the Fourth International Conference on Closed Ecological Systems and Biospherics in 1996. A half dozen presentations shared results from Biosphere 2's two-year closure. Our Russian

Figure 71. Speakers at the Linnean Society meeting, London, 1996. Back row, from left to right: Mark Mills, T. Heflin Jones, Craig Litton, Ray Collins, author, John Allen, Galina Nechitailo, Nicholai Pechurkin, Frank Salisbury, Lydia Somova, Roger Binot, Ganna Meleshko, Alexander Mashinsky, Sergei Zhukov. Front row: Seishiro Kibe, Tsutomu Iwata, Yevgeny Shepelev, Abigail Alling, Phil Dustan, Sally Silverstone, Andrew Brittain. Not in photo: Sir Ghillean Prance, William Dempster, William Challoner.

friends presented their current work as did people working with NASA, the Japanese and European space agencies, and the Ecotron closed ecological system at Imperial College, London. Many of the papers were published in a special issue of the journal *Life Support and Biosphere Science*.[9]

Changes at Biosphere 2

I didn't fully realize how radical and paradigm-shifting Biosphere 2 was until I saw the changes after we left. When Columbia University took control of the facility in 1996, they were primarily interested in studying the impacts of elevated carbon dioxide on the complex biomic systems we had built.

Changes included eliminating people living in Biosphere 2 because they wanted to study a biosphere without humans. Not having people as part of the experiment brings it more in line with the current norms of research. To be objective, scientists need to keep their distance and dispassionately study the biosphere!

Eugene Odum, in a letter to the journal *Science*, made the case for new approaches in science:

> The mission of this venture is not generally understood by the scientific community. The mission of this experiment is not traditional, reductionist, discipline-oriented science, but a new, more holistic level of ecosystem science that has been called "biospherics." Biosphere 2 is as much a human experiment as a scientific one. When you consider that nothing on the scale of Biosphere 2 has been attempted before . . . and how little we really know about how our Biosphere 1 (Earth) works, a measure of success will have been achieved if the biospherians come out alive and healthy this fall after the 2-year isolation. Certainly the experiment will have improved our understanding of human-biosphere interrelations and helped answer the question of how much natural environment must be preserved for life support, and it will have provided a basis for improving the design next time around.[10]

Designing a technosphere or agriculture compatible with a biosphere also didn't qualify as a worthy research topic. The agricultural system was removed though the impacts of global climate change and elevated levels of greenhouse gases on food crops are huge concerns. Where the farm once thrived, they tested the impact of varying CO_2 levels on stands of cotton-wood trees.[11]

Then dividers (large plastic sheets) were installed between biomes so they could study each separately, reducing interactions between them. To facilitate these studies, Biosphere 2's life as a closed ecological system ended. Instead, the facility operated in "flow-through" mode; carbon dioxide was injected at desired levels and the atmosphere was vented accordingly.[12] I privately mourned "the destruction by reduction of Biosphere 2." But important research would continue to come from Biosphere 2, and I needed to reconcile myself to the facility's new incarnation.

Biospheric Politics

Post-closure Biosphere 2 was, for a few years, a surreal place of nasty scientific politics.

When I visited in 1996, the only photographs on display were of members of the 1994 second closure experiment. The only mentions of the first closure emphasized lack of food, oxygen, and ecological problems. No accomplishments or discoveries were acknowledged or any goals apart from "space colonies." A couple years later, there were some photos of the first team, but just those of the four who backed a change of management! It reminded me of how the Kremlin used to airbrush out of photos cosmonauts who died in space or ideologically disgraced former members of their Central Committee.

It got stranger, almost beyond what you'd dream up as a dark comedy satire. Biosphere 2 was opened for inside tours by visitors and, at least for a time in 1999, you could walk around without a tour guide. I was stunned to wander into "the animal bay," a wing on the ground floor that once housed goats, chickens and pigs. There was an "educational exhibit" on how the automobile industry was responding to global climate change and other environmental threats. In the center was a gleaming car!

Mr. Toyoda almost got his wish to see a car inside Biosphere 2. But it was not his company's, but a Volvo, since they had given Columbia millions in scholarship money. The company was even allowed to help develop course materials for Columbia's Earth semester students at Biosphere 2. This arrangement aroused controversy, as you might expect, since it was so obviously a display of corporate influence.[13]

Still reeling from Biosphere 2 becoming an ad placement for the car industry, I wandered down the habitat hallway. I peered in through a new display window to the room I'd lived in for two years. The sign outside said it was the room of Bernd Zabel, a biospherian from the second six-month closure. I was not mentioned.

At the Biosphere 2 bookstore, they told me the first book I co-authored about Biosphere 2 (with Gaie and Sally), *Life under Glass: The Inside Story of Biosphere 2*, was out of print. I learned later it had been recalled from every U.S. bookstore, some two thousand copies, at the expense of Biosphere 2's new management and were stored (quarantined?) a few hundred yards from Biosphere 2. I had to threaten legal action to get control of the book to get it back into distribution and let foreign translations go forward. Hard to rewrite history when there is a firsthand account out there written by three of the crew before we exited Biosphere 2!

Columbia issued news releases that mirrored our own, about the exciting potential that Biosphere 2 offered. But while proudly saying that the engineering of Biosphere 2 was "world class," they asserted there was no worthwhile science being done before they arrived.[14] There was no mention of John Allen or of the actual goals and results of the earlier closures. We were simply dismissed as space-nuts, '60s radicals, and/or actors from New Mexico. The bizarre politics and attempt to rewrite Biosphere 2's history helped motivate me to write this book.

My undergraduate degree from Dartmouth College included a philosophy major. So, I know a fallacious *ad hominem* argument when I see one. Personal attacks do not invalidate someone's work or reasoning. But in today's tabloid news culture, personal attacks often substitute for real discussion. Was this a contributing factor in my decision to get a PhD after Biosphere 2? Now my enemies have to call me "Dr. Nelson."

How did we create this engineering marvel, not to mention the biodiverse biomes perfect for studying the impacts of global warming? It was the reverse of the ET rumor: perhaps Biosphere 2 had simply arrived like a UFO and landed in Arizona. Or, as Tony Burgess, put it, Biosphere 2 grew like a mushroom in the Arizona desert until it was discovered.[15]

Space Life Support

In bioregenerative life support, Biosphere 2 remains a road opened but not taken. Part of the Biosphere 2 creative team self-financed and built another closed system, the Laboratory Biosphere. Under a joint venture of Global Ecotechnics Corporation, the Institute of Ecotechnics, and the Biosphere Foundation, we conducted research in it from 2001 to 2007 at our facilities in New Mexico.

The Chinese have taken world leadership in the field with two very advanced facilities in Beijing. One is called Lunar Palace (Permanent Astrobase Life-Support Artificial Closed Ecosystem) at Beihang University. They successfully completed a three-and-a-half-month closure with three people, recycling wastes and growing most of the required food.[16] The other is at the Chinese Astronaut Center. Both teams' goals are very ambitious, sizing up to facilities comparable to Biosphere 2 eventually for extended duration closure experiments. Both groups are well-funded with expectations of benefits for Earth applications as well as opening up space for long-term habitation. The European Space Agency's program, on a more modest scale, called MELISSA (Micro-Ecological Life Support System Alternative) is headquartered at the University of Barcelona.

Scientific Methods

The politics around Biosphere 2 were so intense for a time that when the journal *Ecological Engineering* commissioned four special issues of research at the facility (resulting in a book published by Elsevier in 1999), they had two co-editors from the opposing camps. One editor was Dr. Bruno Marino, the first director of research after Columbia University began managing the facility, and the other was H. T. Odum, who was supportive of the original purposes of the facility.

Odum and Marino pointed out the various uses that Biosphere 2 might serve, noting:

> As a facility, or as a prototype for an experimental ecological facility of the future, the time for large-scale experimental systems such as Biosphere 2 has come. . . . Because the complexity and costs of such a large scale facility are enormous, perhaps Biosphere 2 could become a national laboratory, operated for all those studying sciences of the Earth.[17]

The endeavor of science is expansive: understanding the material world, nature, and the universe. Students, though, are often taught only one approach: a hypothesis tested in an experiment to determine if it is true or false. That only describes one type of science. It notably does not include observational sciences. Darwin had his insight about natural selection and evolution on an around-the-world expedition where he carefully observed the plants and animals he encountered. Science through observation includes naturalist ecology, Earth systems science, geology, and astronomy. These and other sciences can't always reduce what they are studying to a few discrete variables or make experiments.

Biosphere 2 was built using integrative, holistic system approaches and a lot of detailed, analytic science. Some of the controversy the early closure experiments aroused was due to the unfortunate chasm between analytic and systems scientific approaches. At present, the far stronger faction is the analytic or reductionist scientists.[18] This approach has yielded enormous advances through its study of nature at very small, fundamental levels. But many question whether this approach alone is adequate for ecology, where complexity is extreme and multiple vectors and scales of organization are inherent. Analytic scientists tend to be dismissive of systems approaches. They often don't recognize that multiple approaches to science are needed with each yielding their own insights and understanding.

Odum noted in his paper, "Scales of Ecological Engineering":

> When journalists asked establishment scientists, most of whom were small scale (chemists, biologists, population ecologists), they got back the small scale dogma that system-scale experiments are not science . . . Some people

recommended Biosphere 2 be used as they have used growth chambers for 60 years to study small things with many replications, relate trees to carbon dioxide, study species dynamics etc. How do you explain to people whose lives have been dedicated to organismic or population scale that what is more important on an ecological mesocosm scale is the whole self-organizing process? The real world of Biosphere 1 and Biosphere 2 has several scales of size all interacting together . . . A priori, all scales of science may be of equal importance, but there have been large research funds for the small scale and very little for experiments at a large enough scale to be relevant to the global atmosphere . . . There is no sure way to test theories and models of mesoscale self-organization except by seeding and running mesoscale systems. Science at one scale cannot validate that at the next scale.[19]

The bias toward analytic, small-scale science and against holistic studies limits science's relevance to global biospheric challenges. Add to this the difficulty of implementing truly interdisciplinary research needed to comprehend the Earth. Dr. James Lovelock, co-originator of the Gaia hypothesis, called it *academic apartheid*: institutional barriers that prevent scientists of different disciplines working together.[20]

To design Biosphere 2, we brought scientists and engineers together as well as researchers from many different scientific fields. This needs to be part of a new paradigm in science to make it of more use in assisting our transition to a healthy relationship with the biosphere.

John Allen, in his talk at the Linnean Society conference, made the argument that complex ecological systems need to be studied with a variety of approaches.

Four basic ways uneasily co-exist in science to deal with understanding complex systems: [1] prolonged naturalist observation, description of observed regularities and classification of parts . . . [2] analyzing component parts of the object of study, formulating restricted hypotheses, and then, holding all else other than the chosen part as constant as possible, measure changes produced by measured impacts . . . [3] accept complexity as an irreducible element, and then to search for the organized structure that enables us to examine the entity as a whole, to ascertain its specific laws or regularities . . . [4] Put into an operating model a synthesis of these three approaches, together with test principles of engineering, to test the validity of the existent thinking's predictive powers, and to provide a fecund base for new observations. This full interplay of observation, analysis and structuring to make a working apparatus in order to test and extend our knowledge of biospherics is the approach we used to create Biosphere 2. This interplay of all four scientific approaches is required to study Earth's biosphere, the most complex entity yet encountered.[21]

Personally, I found it painful to visit Biosphere 2 in the late 1990s. Maybe I was still too emotionally bonded with its life systems. I could feel Biosphere 2 was suffering from a lack of tender loving care. It looked forlorn and neglected. Some of the biomes looked distressed. As Roy noted when he looked back at our extraordinary experience inside:

> The crew members were able to prevent ecological catastrophe in a style known as 'adaptive management.' The 2-year experience inside Biosphere 2 fitted rather well into Odum's definition of 'ecological engineering' as 'light management that joins human design and environmental self-design so that they are mutually symbiotic' . . . the crewmembers looked at Biosphere 2 as 'our baby' . . . When I exited the enclosure in 1993, Biosphere 2 was a luxuriant paradise of plants, albeit with extinction of some species, a phenomenon commonly observed with self-organizing islands.[22]

I was struck by the symbolism that one experiment in "flow-through" era Biosphere 2 was a test of how long piñon pine trees took to die in containers in the facility. This may be an indication of unpleasant news ahead in a climate-change-warmed U.S. southwest—more droughts and less rain. The trees were not watered and died faster in the hotter, drier desert and slower in the cooler, moister rainforest. Scientists took careful measurements and studied the trees as they died.[23]

Rebecca Reider tracked down John Allen and others from the original team because she was curious why the early history was being concealed. Her book on Biosphere 2 discusses how "mainstream science" (Columbia University) received a free pass with little critical media scrutiny from the moment they took over the facility. The originators of Biosphere 2 were perceived as outsiders (despite our esteemed consulting scientists and institutions) and resented for going past traditional barriers and paradigms of establishment science.

Reider succinctly analyzed how Biosphere 2 broke four unwritten taboos of media and popular views of science.

> "Science" could be performed only by official scientists, only the right high priests could interpret nature for everyone else. . . . . "Science" was separate from art (and the thinking mind was separate from the emotional heart) . . . "Science" required some neat intellectual boundary between humans and nature; it did not necessarily involve humans learning to live with the world around them. Finally, "science" must follow a specific method: think up a hypothesis, test it and get some numbers to prove you were right.[24]

Some of these notions go back deep in the evolution of mainstream modern science, like the divorce of "experience" from "experiment." There were a few voices, like H. T. Odum, who weren't fazed by the controversies:

> The original management of Biosphere 2 was regarded by many scientists as untrained for lack of scientific degrees, even though they had engaged in a preparatory study program for a decade, interacting with the international community of scientists including the Russians involved with closed systems. The history of science has many examples where people of atypical backgrounds open science in new directions, in this case implementing mesocosm organization and ecological engineering with fresh hypotheses.[25]

The antagonism has thankfully finally died down. Columbia University left in 2003. The University of Arizona stepped in a few years later to take over management and evolution of the facility with generous financial support from Edward Bass. Now there is real acknowledgment that Biosphere 2 accomplished important ecological research during its early years.

Missing Data

It is my hope that all the data from the early closure experiments are converted to modern digital format and made accessible via Internet to the international scientific community. But where is the data? The University of Arizona Biosphere 2 directors have assured colleagues and me that they don't have the historic data.

It's shameful to simply accept that these important data are gone. Only a very small percentage of the data from the closure experiment era has been analyzed and published so far. The full data from the food crops and biomes, which developed for four years under conditions of elevated CO_2, should be extremely interesting for its insights into community level responses, biodiversity changes, and ecological self-organization. The detailed evolution of the biomes correlated with environmental sensor data is unprecedented for all kinds of ecological investigations. Most climate change studies are conducted on one or a few plants, in small experimental chambers (phytotrons), though there are a few studies of larger communities.[26]

The Biosphere 2 data about the metabolism of an intensively monitored biospheric mesocosm remains a remarkable and unique but thus far mostly inaccessible resource. I fear that some or most of it will be or is already lost, due to neglect, entropy, or a side effect of the politics and contempt once shown toward the early work at Biosphere 2. The data would reflect well on everyone connected with Biosphere 2 and would contribute to present work with the facility, which makes its disappearance hard to understand.

It would be wonderful if a historical archive could be established to preserve and make available these data and the documentary videos (many also currently unavailable) showing how the facility was created. The University of Arizona science library would be an obvious candidate as the most appropriate institution.

The full harvest of what was learned at Biosphere 2 from its yet-untapped wealth of data can still occur. Many scientists are eager to access the full data set from this unique and extraordinary experiment.

When I go to space conferences on life support (I chair sessions on "Closed Ecological Systems for Earth and Space Applications" and "Innovative Approaches to Space Habitation" for COSPAR, the International Committee on Space Research) and speak to knowledgeable ecologists, Biosphere 2 is treated as a landmark pioneering project. A few years ago, Bill Dempster and I gave keynote speeches at a conference on bioregenerative life support at the Lunar Palace facility in Beijing. We were told that our work is very famous there. Every child in China studies Biosphere 2 in school ecology textbooks.

In fact, it's extraordinary how many people, in the most diverse regions, remember Biosphere 2 as an amazing project. It's just that few people know much beyond the images they saw. That also fired my desire to more fully tell the story of Biosphere 2—its goals and results—and its relevance to our planetary biospheric challenges. It's a very important story that needs to be told and digested and hopefully built upon.

Important Research in Biosphere 2's New Eras

Some important research was accomplished between 1996 and 2003 when Columbia University ran Biosphere 2 and since 2007 under the University of Arizona, though the facility is no longer a closed ecological system nor a total systems biospheric laboratory.

Columbia built dormitories for students and researchers, and semesters at the facility attracted students from more than two hundred international and U.S. universities. The creators of Biosphere 2 had dreamed of a biosphere university at the site someday. Columbia started and the University of Arizona is continuing to make this a reality with their educational programs for students and the public.

The early work on the response of stands of cottonwood trees to different levels of elevated CO_2 yielded important results. Plant physiologist Ramesh Murthy, a Columbia University professor involved in the research, made the case for doing this research inside Biosphere 2.

> We can measure what goes in and what leaves the chamber. We can use these cottonwoods to develop models to predict where carbon dioxide and other [atmospheric] compounds go. Unlike in a real forest, we can control the conditions here. That means we can do experiments that cannot be done in real forests or in normal laboratories because they lack controls.[27]

These studies did reveal the surprises that studies of whole stands of trees rather than individual trees or even individual leaves make possible.[28]

Then even more significant research results came from studies of the Biosphere 2 biomic areas taking advantage of their biodiversity and the manipulation of vectors possible. Much of this research brilliantly combined both analytic and whole system science. They did not lose the forest for focusing on a tree, or even a leaf from a tree. These studies and the ones currently underway by the University of Arizona are showing that Biosphere 2 can indeed help make ecology more of an experimental science.

Several years of research on Biosphere 2's ocean demonstrated the devastating impacts from elevated atmospheric CO_2 and acidification that will result from continued global climate change. The coral reef was studied at 200 ppm, 350 ppm, 700 ppm, and 1,200 ppm of CO_2. Corals grew twice as fast at the lowest levels (probably similar to those of Earth during the last ice age) compared to the 350 ppm levels in the 1990s Earth atmosphere. At 700 ppm, a level which may be reached by 2100 or earlier, coral growth declined another 20 to 40 percent, which might make coral reefs far more susceptible to other degradations. At 1,200 ppm, coral growth declined 90 percent.[29]

Frank Press, former president of the National Academy of Sciences, described these interactions between atmosphere and ocean, taking advantage of the highly controllable ocean mesocosm of Biosphere 2, as the "first unequivocal experimental confirmation of the human impact on the planet."[30]

Similarly, studies in Biosphere 2's terrestrial biomes showed that vegetation reached a saturation point with elevated CO_2 beyond which they are unable to uptake more. Striking differences between the Biosphere 2 rainforest and desert biomes in their whole system responses "illustrate the importance of large-scale experimental research in the study of complex global change issues."[31]

Another set of experiments critically examined the assumption that rainforests will continue to absorb CO_2 as atmospheric concentrations increase. Researchers controlled CO_2 levels in the Biosphere 2 rainforest so photosynthesis and respiration could be studied at 400 ppm, 700 ppm, and 1,200 ppm over two years. Results showed that simply measuring leaves, as is commonly done, greatly overestimates the forest's ability to sequester carbon dioxide. The entire ecosystem behaves differently. The research showed that, even in a couple of decades, CO_2 uptake by tropical forests will begin to decline. That slow-down will accelerate if atmospheric levels continue to rise.[32]

After Columbia University left the biosphere, it looked for a while that the facility might be torn down for high-end housing development. Truth is stranger than fiction: they were to be called "Biosphere Estates." Since, in a sense, the one-hundred-year planned Biosphere 2 experiment is still running, the facility is still teaching us lessons. This prospect offered a bleak commentary on our global ecological crisis. I could see the headlines: BIO-SPHERE CONSIDERED AN EYE-SORE AND INCONVENIENCE, BIOSPHERE DEMOLISHED TO IMPROVE VIEWS FROM HOMES FOR THE WEALTHY. The headline that actually ran in the *New York Times* was even more telling: SPRAWL OUTRUNS ARIZONA'S BIOSPHERE.[33]

University of Arizona Research at Biosphere 2

The University of Arizona (UA) courageously stepped in to ensure that Biosphere 2 would continue to be a place of teaching and learning. They have been managing the facility since 2007, developing new programs for research and education. In 2011, the university assumed full ownership. Edward Bass again demonstrated his support of the facility through his substantial endowment and contributions to the University of Arizona as he had assisted Columbia's management earlier.

Biosphere 2 had entered a new era.

UA developed a robust research program, utilizing some of the existing biomes and developing ambitious new programs.

The centerpiece is the creation of the Landscape Evolution Observatory (LEO). Occupying the space where agriculture was studied, this project is the world's largest and most interdisciplinary Earth sciences experiment. Three replicates of artificial landscapes will permit studies of soil, topology, and living communities in their evolution. LEO will provide insights into global climate change effects on water resources and arid ecosystems. An impressive array of 1,800 sensors and sampling points will enable precise monitoring as the systems mature.[34]

The Biosphere 2 rainforest continues to yield important scientific results. Drought, higher temperature, and CO_2 experiments, reflecting expected outside changes under climate change, will test how the rainforest as a whole and individual trees change their water cycling and carbon dynamics. These studies are compared to those in natural rainforests, such as those of Amazonia in Brazil and Peru.[35]

UA is planning a transformation of the Biosphere 2 ocean to one modeled on the Desert Sea (the Sea of Cortez, also called the Gulf of California). The original coral reef suffered from neglect in the years between Columbia leaving and UA's tenure. The Desert Sea may be an easier biome to maintain, and its rich biodiversity includes subtropical, rather than tropical, coral reefs. The new ocean design will include subtidal and intertidal zones. Preparatory work now underway includes aligning current ocean chemistry with that of the natural ocean and removing excessive algae.[36]

Biosphere 2 and its associated dormitory and other buildings are being studied to test new ways of lowering energy costs and water use. This includes testing of solar panels on challenging high-slope terrains, developing new types of microgrids to maximize the efficiency of renewable energy generators, and installing green roofs to reduce energy costs of buildings and cities.[37]

UA's trace gas laboratory is developing new tools to measure greenhouse gases. Research on the critical zone, the place where the Earth's surface is shaped, is being done inside Biosphere 2 biomes and the LEO and outside on a nearby area, which is part of the national network of critical zone observatories.[38]

The University of Arizona proudly supports a substantial artist in residence program at Biosphere 2. The resulting art, photographs, videos, and performance installations expand the impact of the facility as sculpture, stunning architecture, icon, metaphor, myth, and mystery.

People told us when we built Biosphere 2 that we were fifty years ahead of its time. That may still be true, but now, more than two decades later, people realize how much we need its research. Changes in ecological awareness have been rapid in the years since the facility was designed and operated. But so has the assault on the global biosphere. It is no overstatement to say that we are in a race and that the stakes are unprecedented for our health and that of our biosphere.

It is gratifying to see Biosphere 2 flourishing under the management of the University of Arizona. Seeing the research being done there, and the continuing inspiration the magnificent structure gives to visitors, Biosphere 2 continues to play its part in waking us up.

Conclusion

A Time to Tear Down, a Time to Rebuild

Momentous Times

WE ARE COMING to know Earth's biosphere at a very troubling time. The Industrial Revolution, in less than three hundred years, has intensified and spread throughout the world spawning unintended consequences everywhere.

Some actions are obviously needed. We need to slow down, halt, and eventually reverse what greenhouse gases are doing to our atmosphere. Otherwise, the impact on ecosystems, acidification and warming of our oceans, and increase in extreme weather events, on top of other human impacts on our biosphere, make for a truly frightening and even unpredictable near and long-term future.

Our future is not predetermined. It's certainly not too late. We can't wait for the future; we must create it. All is not gloom and doom. Many incipient positive trends and changes can reverse the threats to our biosphere's future—and ultimately our future. Perhaps more important than technical advances are changes of consciousness, how we understand our relationship to nature.

The Paradigms, They Are a Changing

Thomas Kuhn, in *The Structure of Scientific Revolutions*, notes that real break-throughs don't proceed by a logical, step-by-step improvement of existing ways of understanding the world. Breakthroughs come in episodic bursts when new paradigms replace inadequate older ones.[1] Paradigms are models, "a set of assumptions, concepts, values, and practices that constitutes a way of viewing reality for the community that shares them."[2]

The dominant modern paradigms about our biospheric relationship are dysfunctional and false. "Not in my backyard" has expanded as we under-stand the entire Earth is our home. Quebec opponents of fracking say *"ni ici, ni ailleurs!"*—not here, not anywhere![3] We all share an atmosphere and a global water cycle, so what happens in one region affects us all.

New paradigms are urgently needed because of the power of industry, increasing population, demand for better living conditions, and a techno-sphere that causes ecological and human health crises impossible to ignore. The Millennium Ecosystem Assessment found that 60 percent of global ecosystem services are degraded or used beyond replacement rates.[4] E. O. Wilson, the Harvard biologist, has boldly called for half of the planet's ecosystems to be left untouched and, where needed, restored to health so that we preserve biodiversity and don't risk damaging critical biospheric functions.[5]

Growing Up in the Anthropocene

Humanity, in these early years of the Anthropocene, behaves like a child with new toys. Little did those who turned to coal, then oil and natural gas, to power our industry and lives understand they were doing more than freeing us from the limitations of water wheels, windmills, and sails. Tall smokestacks spewing black by-products of those fossil fuels were originally seen as symbols of progress and wealth.

Max Planck, who revolutionized physics by inventing quantum theory, noted: "A new scientific truth does not triumph by convincing its opponents and making them see the light, but rather because its opponents eventually die, and a new generation grows up that is familiar with it."[6]

Younger people around the world understand far better than their elders that new approaches are needed. They don't take as givens that our technosphere must rely on nonrenewable and ecologically damaging resources, that our food must be grown with chemical fertilizers and pesticides, or that we have to put up with polluted air and water. The switch to wind and solar power is proceeding fairly rapidly in countries that have made the decision. Progress is far slower in others where the powerful interests of giant companies and their government allies strenuously deny realities like climate change and exaggerate the economic costs and disruptions of a shift to nonpolluting renewable sources of energy.

The biospheric paradigm shift is up against powerful ideologies deriving from "neoliberalism" and free market economics.[7] Their premise that market forces will solve any environmental problem has not worked, and there is no reason to believe it will in the future.

The paradigm shift is exemplified by terms signaling a new relationship between humans and our biosphere. Intergenerational justice emphasizes meeting current needs without damaging the capacity of future generations to fulfill their needs.[8] It's encouraging that sustainable development is widely used despite confusion about what it means. *Sustainable* was not a word people said or thought about much in the early 1990s. Otherwise, we would have referred to it in explaining Biosphere 2's approaches to growing food, recycling water, and designing a life-supporting technosphere.

Though we may not precisely know what is sustainable, it's clear that many aspects of "business as usual" are *not* sustainable and change our world for the worse. We need new economic paradigms since conventional economics still treats shared resources and the environment as an externality— outside of and therefore not included in economic analyses. Unless polluters pay for clean-up and damage to the biosphere, there are no economic incentives for them to change.

Ecological economics emphasizes the interdependence and co-evolution of ecological systems and human economics. This expanded perspective values "natural capital," the health of our biosphere, which contributes so many critical free services and maintains its life support capacities.[9]

Real Measures of Wealth and Well-Being

Real wealth and human well-being are not measured by gross national product. GNP grows with every accident, illness, and pollution-causing manufacture since money changes hands. Real wealth, not money, is our ability to lead fulfilling lives while regenerating our world. Every human being on Earth can share this type of real wealth since we live on an abundant planet with which we are just learning to integrate successfully. New wellness and wealth measures include the OECD's Index of Better Life[10] and the UN and World Bank's Inclusive Wealth, expanding the current economic matrix to include natural capital.[11] The king of Bhutan declared he's more interested in his country's gross national *happiness* than in their gross national product.[12]

It's a false and dangerous assumption that we have to choose between healthy economics and a healthy environment and biosphere. The groundbreaking *Limits to Growth* pointed out the environmental impossibility of exponential growth in a finite system (Earth's biosphere).[13] In *Beyond the Limits*, it is argued that though the world has reached "overshoot" in some areas it is possible to provide for everyone's well-being while restoring environmental health. To do this will require maturity, compassion, and justice. Increasing our quality of life must supplant physical expansion of production.[14]

The words ecology and economics derive from the ancient Greek, *oikos*, meaning house or household. In the new paradigm, they are complementary. Any economic system that destroys its house, its natural capital, is self-destructive and therefore short-lived. It is difficult to translate some aspects of the biosphere into monetary terms; some are irreplaceable and deserving of life for their own sake.

Estimates of what nature does for us for free range from $33 to $140 trillion dollars per year. This makes nature roughly equal to or three times larger than conventionally determined world GNP. These studies can help change our worldview. They highlight previously ignored natural assets and services that underlie human well-being and underscore our interdependence with the health of the biosphere.[15]

Mistreatment of the biosphere is already producing dire consequences that will only increase if we don't change our behavior. In Biosphere 2, I watched extensive coverage of the Sahel famine of the early 1990s without any mention of the environmental degradation largely responsible. Similarly, refugee crises are rarely linked to underlying causes. Years of severe drought and water shortages linked to climate change contributed to the Syrian civil war.[16] Environmental refugees now equal those forced to flee political, ethnic, and religious persecution.[17] The twenty-five million environmental refugees of 1995 may grow to fifty million by 2020 and, if climate change is not abated, there may be 150 million environmental refugees by 2050.[18]

Redesigning Our Technosphere

Technologies currently used are often neither efficient nor obligatory. The Industrial Revolution greatly increased living standards and life expectancy for much of the world. But driven by an economics and paradigm where natural resources were considered abundant if not infinite, many industries chose relatively crude approaches. Pollution and waste were not factored in, as they were discharged into our commons.

I ridiculed the engineering slogan "the solution to pollution is dilution," thinking it was long dead. Fellow graduate students told me that down the hall from our systems ecological engineering corner in the environmental engineering sciences building, professors were still teaching using that slogan. Have a pollution problem? There's an easy fix, build the smokestacks higher and extend the sewage pipes a few more miles into the sea. This puts off redesigning our technosphere so it doesn't pollute to begin with.

Many engineers and scientists think resources can be used five to ten times more efficiently, resulting in an 80 to 90 percent resource reduction.[19] This would dramatically change the relationship between use of natural resources and production of goods and services.

Just as the engineers rose to the challenge of designing a life-supporting technosphere in Biosphere 2, this can be done in our global biosphere. Already, shifts to a "circular economy" emphasizing zero-waste and zero-emission technologies are showing economic benefits. The paradigm shift of designing industrial ecosystems, where wastes (any by-products) from one manufacturing process is an input to another also is being put into practice.[20]

Agribusiness contentions that small-scale, organic, and more natural food production approaches cannot feed the world are increasingly being questioned.

The UN Conference on Trade and Development called for an "ecological intensification" paradigm shift to replace the Green Revolution, with its energy and capital-intensive approaches. World agriculture problems will not be fixed by just tweaking current production methods. Seventy percent of the world's people suffering from malnutrition are poor farmers and farm workers. Empowering them to integrate new approaches, inspired by traditional practice, including agro-forestry, integrating crops and animal raising, incorporation of wild vegetation areas, and better use of organic and conventional fertilizers (including closed nutrient cycles like was practiced in Biosphere 2), is seen as the best approach to giving food security to all the world.[21]

It is imperative that farming families and communities gain societal respect and earn enough money to raise their living standards and educational access, especially for women. This is also vital for reducing population growth.[22]

The strengthening of healthy, ecologically intensive farming mirrors the rapidly growing demand for chemical-free and local produce. Though organic agriculture worldwide currently is 1 percent of total farmland, its growth is rapid. Farmers see the high costs of chemical farming and are shifting to minimum tillage and more natural farming approaches. Organic certification systems operate in more than 160 countries; in ten of these countries, organic production increases by more than 10 percent per year.[23]

Taking Care of Everyone aboard Spaceship Earth

At the dawning of the ecological movement, Buckminster Fuller powerfully visualized the Earth and its biosphere as "Spaceship Earth." He foresaw "ephemeralization"—doing more with less—as crucial along with comprehensive, anticipatory design, intelligent retooling of our technosphere, and discarding outmoded thinking in favor of new paradigms. Though the Earth is superbly organized by life, we humans have yet to figure out how to manage ourselves. Spaceship Earth came with neither an operating manual nor a warranty. In our ever more connected world we must act so that all benefit. Fuller predicted, "It has to be everybody or nobody."[24]

If we rise to our challenge, the developed and wealthier countries will help the poorer ones use ecologically appropriate and more carbon-neutral renewable energies to improve their living standards. This is our climate debt—it is unfair for poorer countries to suffer because of climate change they did not cause.[25] Nor can they be asked to forego development to help biospheric health.

Reducing CO_2 release associated with economic output is proceeding faster than expected driven by sharp declines in wind and solar electric costs. 2014 marked the first time the world economy grew without an increase in greenhouse gas emissions. This is far from the sizeable declines needed, but it is a hopeful trend.[26]

Some new solar installations are economically competitive with fossil fuel and may be fully competitive in a decade.[27] That developing, poorer countries might largely bypass fossil-fuel-driven electrical generation is now looking possible even by bottom line calculations.[28] This would mirror how many such countries went directly to cell phones bypassing more expensive and resource-consumptive landlines. A carbon tax would more correctly price our current use of fossil fuels since its damage to the biosphere is still an unaccounted "externality."

We need a Marshall Plan scale aid program to help the poorer, developing world make the transition to a regenerative economy. It is a question of priorities: many trillions of dollars were found to shore up Wall Street megabanks after the 2008 bailout. Annual world military expenditures in 2012 were higher than $1.7 trillion, some 2.5 percent of world GNP.[29]

The 2006 Stern Report to the UK government predicted huge economic losses if climate change is unchecked. They concluded spending $1 to $2 trillion per year (roughly 1 to 2 percent of world GDP), including transfers from the wealthy countries to developing ones, makes sense even from a purely economic analysis.[30] The investment in making the transition to a post-fossil-fuel economy is urgently needed soon. The longer we wait, the worse the damage and the more expensive investments will need to be. But when societies decide that something is really important, it can be done.[31]

Nature Deficit Disorder

The shift to a regenerative and healing relationship with our biosphere will improve our well-being. "Nature deficit disorder" in our increasingly urbanized world perhaps explains why so many people find their lives unsatisfying.[32] Something is missing. This divorce from nature may contribute to many mental and physical disorders. The expansion of parks, green areas, and food production in cities is a win-win-win situation. City parks and urban agriculture make the pleasures of nature more accessible, improve air quality, and reduce greenhouse gases as well as delivering fresher food.

I presented a paper on constructed wetlands ecological sewage treatment at a UN conference titled "Cities as Sustainable Ecosystems" at Murdoch University in Perth, Australia. The workshop helped me grasp the power of cities as major drivers of the world economy with their food, water, energy, transportation, and communications needs. While occupying just 2 percent of Earth's land, cities are responsible for 78 percent of human greenhouse emissions, 76 percent of wood use, and 60 percent of our water use.[33]

Greening our cities through environmentally friendly buildings, renewable energy generation, better public transport, in-situ and nearby food production, and ecological methods of sewage treatment are benefits in themselves while also decreasing negative environmental impacts.

When a reviewer of a scientific paper I wrote challenged me to substantiate that living in a beautiful green world was psychologically beneficial for the biospherians, I discovered much evidence supporting this common-sense knowledge.[34] For example, hortitherapy is an emerging field, applying what humans have always known—the health and happiness of caring for green plants and spending time in gardens.[35] Healing and greening our world will help us heal ourselves.

New Stories, New Myths

It's important to not heed those ready to give up the fight for a better future, as though we've passed some irrevocable tipping point. Artists and writers, poets and dreamers among us have a critical role to play. We are in great need of new scripts, new storylines, new epics, new mythologies, and teaching stories for humanity as we forge a renewed respect and moral compass for our behavior toward the natural world in the Anthropocene.

Many of the old myths deeply embedded in our cultural consciousness have sobering tales to tell: expulsion from the Garden of Eden, Gilgamesh cutting down the sacred forest, the ever-receding "golden age," paradises gained but lost.

William Burroughs spoke at the 1980 IE Planet Earth conference. After listening to a weekend of reports on global challenges, he rose to sardonically say that he'd found a title for his next book—*Earth: The Place of Dead Roads*. Burroughs observed that we need to replace the tired old scripts with a new mythology where we will again have "heroes and villains judged by their intentions toward the planet."[36]

Outdated mythic baggage, narrow bounds on science, and economics that exclude human and natural values help feed the strange separation of humans from their rightful roles as participants in the biosphere. Many people think the environment is something outside of themselves, rather than understanding the truth that we are inseparable from nature, from our biosphere. These old paradigms are wrong, dispiriting, and harmful.

The first step on the Buddhist path to awakening (Buddha means the awakened one) is Right Understanding, Right Vision. You have to change the way you think. You have to see the world the way it is, not as we want to or are conditioned to see it.[37] The future need not be a march into collapse as extrapolated from our current assault on the biosphere. Beware self-fulfilling prophecies; they are delusional.

We need to awake to our glorious reality; virtually everything I've said about the biospherians in Biosphere 2 is equally applicable for us living in Earth's biosphere.

I gave a talk on ecotechnics and Biosphere 2 at the Dallas County Community College District Sustainability Summit. Improvising at the end, looking out at a sea of young faces, I found myself saying: "We're all looking for our soul mate, but we already have one, it's our biosphere." Evolutionary biologist Stephen Jay Gould made a similar point: "We cannot win this battle to save species and environments without forging an emotional bond between ourselves and nature as well—for we will not fight to save what we do not love."[38]

Changes of Consciousness

Changes in consciousness are crucial to forging new paradigms. The spread of less materialist spiritual traditions, like Buddhism, yoga, and meditation are evidence of change. There is growing appreciation for indigenous and other traditions which treat Earth as sacred. Use of entheogens (sacramental substances) is growing, many of them long used by indigenous peoples in healing and spiritual ceremonies. Mindfulness practices are increasingly employed to treat PTSD and other ailments and to deepen and enrich our experiences.[39]

These all serve as antidotes to the externalized life accumulating "stuff," which modern culture inculcates. Albert Hofmann, chemist and researcher of consciousness-expanding substances, wrote: "Alienation from nature and the loss of the experience of being part of the living creation is the greatest tragedy of our materialistic era. It is the causative reason for ecological devastation and climate change. Therefore I attribute absolute highest importance to consciousness change."[40]

The "Green Christian" movement shows the broad reach of the paradigm change. Environmental Christians interpret the Old Testament passage giving humans "dominion" over nature to mean "stewardship" rather than a license to extract, control, and despoil.[41]

Pope Francis's 2015 encyclical letter, "On Care for Our Common Home," makes an impassioned call for a dramatic change in human thinking and behavior to avoid ecological catastrophe. As spiritual leader of 1.2 billion Catholics worldwide, his forceful stand is significant in its recognition of our profound link and debt to the biosphere.

Our sister [Mother Earth] now cries out to us because of the harm we have inflicted on her. . . . We have come to see ourselves as her lords and masters, entitled to plunder her at will. . . . This is why the Earth herself, burdened and laid waste, is among the most abandoned and maltreated of our poor . . . We have forgotten that we ourselves are dust of the Earth . . . our very bodies are made up of her elements, we breathe her air and we receive life and refreshment from her waters . . . a true ecological approach always becomes a social approach; it must integrate questions of justice in debates on the environment, so as to hear both the cry of the Earth and the cry of the poor . . . distractions constantly dull our consciousness of just how limited and finite our world really is. As a result, "whatever is fragile, like the environment, is defenseless before the interests of a deified market, which become the only rule."[42]

The Dalai Lama of Tibet has said, "Compassion is the radicalism of our times." Compassion for all sentient beings would be a radical change in how humans treat our fellow species and the biosphere.

The Human Experiment

I remain optimistic about the prospects for the "human experiment." Problems humans cause can also be solved by humans. My optimism, in large degree, comes from my Biosphere 2 experience, which taught us every action, however small, is important. If we give up hope, we cease motivating ourselves to act for positive change. This is not fatuous optimism, which ignores the forces and inertia arrayed in support of the existing order of things. Optimism is a yoga, a mental discipline. My optimism seeks more than simply avoiding biospheric collapse.

Changes can come quickly. I first went to the Soviet Union in 1985, thinking it was a powerful empire. Our first Christmas in Biosphere 2, six years later, we watched Gorbachev dissolve the country. In such tumultuous times, with incomparably faster rates of change, it's understandable that some seek an imaginary escape to simpler times, to turn back the flow of history. The present confused time also means there's an opening for new paradigms based on reality to take hold. Clearly seeing that old, established ways of thinking, of doing business, are woefully inadequate is a prelude to the new.

The challenges we biospherians faced and overcame in Biosphere 2 also give me hope. We didn't destroy our wilderness biomes to plant more food. We didn't undermine each other or cause harm to the life inside, our life support system. We continued working flawlessly with each other no matter what. We understood that there were higher values and necessities that united us.

Carl Hodges helped invent the term "biospherian" early in the project; much better than "bionaut" or "econaut." He said a biospherian was not only someone who lives in a biosphere but a person who totally *understands* a biosphere.[43] I doubt one can ever totally understand any biosphere, but I do agree we should call ourselves Earth biospherians by increasing our understanding and shaping our actions to match.

Martin Luther King said, "We may have come in separate ships, but we're all in the same boat now."[44] We understood that in Biosphere 2—and that the boat was our lifeboat. That united us as a task group, a team. I see a growing appreciation around the world that we humans are in a shared lifeboat, regardless of our origins and circumstances. The ultimate spur to our collective human intelligence may be when we realize our shared necessity to live with one another and our global biosphere.

Our aims should go beyond sustainability to regeneration. Many features of the world we live in need to be changed, including poverty and income inequality, racism and sexism, materialism (which measures life by consumption), and both the diversity of human culture and nature everywhere under attack.

We have passed the time of taking from the biosphere without giving back, for economic and technological systems that damage humans while consuming and despoiling what nurtures us all. We must grow out of our suspended development and think and invent afresh so that we have better ways of living in the Anthropocene. There is no universal template. Life-affirming and successful ways of restoring our biosphere and our human well-being and potentialities will be just as diverse as our cultures and we are.

It is high time to take action and begin creating these new futures. We should act so that future generations look back at our time as when the era of destruction began to be replaced by a restoration of ecological and human sanity. We are collectively facing a species IQ test. It is a test of whether humans can show the intelligence, resilience, and adaptability to be a cooperative and creative part of our planetary biosphere—or whether we are headed toward an evolutionary dead-end.

This is the exhilarating—and yes, scary, because the stakes are so great, the obstacles mighty—challenge of our lives, our time in history.

Always remember we have allies. As we so unforgettably learned in Biosphere 2: We are part of the biosphere, body and soul. The biosphere is on our side.

Notes

Preface

1. F. White, *The Overview Effect: Space Exploration and Human Evolution*, 2nd ed. (San Diego: American Institute of Astronautics and Aeronautics, 1998).
2. R. Schweickart, "No Frames, No Boundaries," *Rediscovering the North American Vision*, no. 3 (Summer 1983): 16, http://www.context.org/iclib/ic03/schweick/.

Chapter 1

1. C. Ponnamperuma, ed., *Comparative Planetology* (New York: Academic Press, 1978).
2. M. Nelson, "Bioregenerative Life Support Systems for Space Habitation and Extended Planetary Missions," in *Fundamentals of Space Life Sciences*, ed. S. Churchill (Malabar, FL: Orbit Books, 1997), 315–336.
3. F. Salisbury, J. I. Gitelson, and G. M. Lisovsky, "Bios-3: Siberian Experiments in Bioregenerative Life Support," *Biosciences* 47, no. 9 (1997): 575–585.
4. V. I. Vernadsky, *The Biosphere*, ed. M. A. S. McMenamin, trans. D. B. Langmuir (New York: Springer-Verlag, 1998).
5. J. Lovelock, *Gaia: A New Look at Life on Earth* (Oxford University Press, 1979).
6. M. M. Kamshilov, *Evolution of the Biosphere* (Moscow: Mir Publishers, 1976).
7. J. Beyers and H. T. Odum, *Ecological Microcosms* (New York: Springer, 1991).
8. G. D. Cooke, "Ecology of Space Travel" in *Fundamentals of Ecology*, 3rd ed., ed. E. P. Odum (Philadelphia, Saunders College Publishing, 1971); H. T. Odum,

"Limits of Remote Ecosystems Containing Man," *American Biological Teacher* 25 (1963): 429–443.

9. C. E. Folsome and J. A. Hanson, "The Emergence of Materially Closed System Ecology" in *Ecosystem Theory and Application*, ed. N. Polunin (New York: John Wiley and Sons, Ltd., 1986), 269–288.

10. "Joseph Priestly and the Discovery of Oxygen," American Chemical Society International Historic Chemical Landmarks, http://www.acs.org/content/acs /en/education/whatischemistry/landmarks/josephpriestleyoxygen.html.

11. M. Roach, *Packing for Mars* (New York: W. W. Norton, 2010).

12. L. Margulis and D. Sagan, *Biospheres: From Earth to Space* (Enslow, 1989).

13. D. Sagan, *Biospheres: Reproducing Planet Earth* (New York: Bantam Books, 1990).

14. J. Poynter, *The Human Experiment: Two Years and Twenty Minutes Inside Biosphere 2* (New York: Avalon Publishing Group, 2006).

15. H. Morowitz, *Biosphere 2 Newsletter* 1, no. 2 (1994).

16. L. Mumford, *Technics and Civilization* (New York: Harcourt, Brace and World, 1934).

17. J. P. Allen, *Me and the Biospheres* (Santa Fe, NM: Synergetic Press, 2012).

18. M. Nelson, chap. 3 in *The Wastewater Gardener: Preserving the Planet One Flush at a Time* (Santa Fe, NM: Synergetic Press, 2014), 13–30.

19. J. P. Allen, M. Nelson, and T. P. Snyder, "Institute of Ecotechnics," *The Environmentalist* 4 (1984): 205–218.

Chapter 2

1. J. Poynter, *The Human Experiment: Two Years and Twenty Minutes inside Biosphere 2* (New York: Avalon Publishing Group, 2006).

2. J. P. Allen, "People Challenges in Biospheric Systems for Long-Term Habitation in Remote Areas, Space Stations, Moon, and Mars Expeditions," *Life Support and Biosphere Science* 8 (2002): 67–70.

3. R. Reider, *Dreaming the Biosphere* (Albuquerque: University of New Mexico Press, 2009).

4. M. Decoust, *Odyssey in Two Biospheres*, Biosphere Foundation, http:// biospherefoundation.org/project/odyssey-of-2-biospheres/.

5. Poynter, *The Human Experiment.*

6. A. Alling et al., "Human Factor Observations of the Biosphere 2, 1991–1993, Closed Life Support Human Experiment and Its Application to a Long-Term Manned Mission to Mars" (paper presented, NASA Mars Ecosynthesis workshop, Santa Fe, NM, September 2000), *Life Support and Biosphere Science*, no. 8 (2002): 71–82.

7.   "Consultants, Scientists Conducting Joint Studies Provide Expertise, Extend Importance of Biosphere 2 Experiment," http://www.biospherics.org/bio sphere2/results/6-consultants-scientists-conducting-joint-studies-provide -expertise-extend-importance-of-biosphere-2-experiment/.

8.   M. Nelson, *The Wastewater Gardener: Preserving the Planet One Flush at a Time* (Santa Fe, NM: Synergetic Press, 2014).

9.   R. Reider, *Dreaming the Biosphere.*

10.   Poynter, *The Human Experiment.*

11.   M. Nelson, unpublished introduction to extracts of Biosphere 2 journal, written in 1995.

Chapter 3

1.   "Theory and History of the Noosphere," Foundation for the Law of Time, accessed June 17, 2017, http://lawoftime.org/noosphere/theoryandhistory.html.

2.   C. Schwägrel, *The Anthropocene: The Human Era and How It Shapes Our Planet* (Santa Fe, NM: Synergetic Press, 2015).

3.   Two million visited by 2002, and a hundred thousand visit each year. "Biosphere under the Glass," Astrobiology Magazine, February 28, 2004, http://www.astrobio.net/news-exclusive/biosphere-under-the-glass/.

4.   K. Kelly, *Out of Control: The New Biology of Machines, Social Systems and the Economic World* (New York: Addison Wesley Publishing Company, 1994).

5.   W. F. Dempster, "Tightly Closed Ecological Systems Reveal Atmospheric Subtleties—Experience from Biosphere 2," *Advances in Space Research* 42, no. 12 (2009): 1,951–1,956.

6.   W. F. Dempster, "Methods for Measurement and Control of Leakage in CELSS and Their Application in the Biosphere 2 Facility," *Advances in Space Research* 14, no. 11 (1994): 331–335.

7.   W. F. Dempster, "Biosphere 2: Design Approaches to Redundancy and Back-Up," *Society of Automotive Engineers Technical Paper* no. 911328.7, Society of Automotive Engineers Twenty-First International Conference on Environmental Systems, San Francisco, 1991.

8.   "Earth Loses 50,000 Tonnes of Mass Every Year," SciTech Daily, February 5, 2012, http://scitechdaily.com/earth-loses-50000-tonnes-of-mass-every -year/.

9.   I. Urbina, "Think Those Chemicals Have Been Tested?", *New York Times*, April 13, 2013, http://www.nytimes.com/2013/04/14/sunday-review/think -those-chemicals-have-been-tested.html.

10.   "The Massachusetts Precautionary Principle Program," Science and Environmental Health Network, www.sehn.org/pppra.html.

11. "Laboratory Biosphere," Global Ecotechnics, accessed November 10, 2016, http://www.globalecotechnics.com/projects/laboratory-biosphere/; M. Nelson, J. P. Allen, and W. F. Dempster, "Modular Biospheres: A New Platform for Education and Research," *Advances in Space Research* 41, no. 5 (2008): 787–797.

12. "The Massachusetts Precautionary Principle Program."

13. H. Morowitz et al., "Closure as a Scientific Concept and Its Application to Ecosystem Ecology and the Science of the Biosphere," *Advances in Space Research* 36, no. 7 (2005): 1,305–1,311.

Chapter 4

1. M. Nelson et al., "Atmospheric Dynamics and Bioregenerative Technologies in a Soil-Based Ecological Life Support System: Initial Results from Biosphere 2," *Advances in Space Research* 14, no. 11 (1994): 417–426.

2. M. Nelson et al., "Earth Applications of Closed Ecological Systems: Relevance to the Development of Sustainability in Our Global Biosphere," *Advances in Space Research* 31, no. 7 (2003): 1,649–1,656.

3. J. P. Allen, *Biosphere 2: The Human Experiment* (New York: Penguin, 1991).

4. A. Alling et al., "Human Factor Observations of the Biosphere 2, 1991–1993, Closed Life Support Human Experiment and Its Application to a Long-Term Manned Mission to Mars" (paper presented at NASA Mars Ecosynthesis Workshop, Santa Fe, NM, September 2000), *Life Support and Biosphere Science* 8 (2002): 71–82.

5. W. J. Rippstein and H. J. Schneider, "Toxicological Aspects of the Skylab Program," in *Biomedical Results from Skylab*, eds. R. S. Johnston and L. F. Dietlin (Washington, DC: U.S. Government Printing Office, 1977); A. Nicogossian and J. F. Parker, *Space Physiology and Medicine* (Washington, DC: U.S. Government Printing Office, 1982).

6. R. M. Hord, *Handbook of Space Technology: Status and Projections* (Boca Raton, FL: CRC Press, 1985).

7. R. Reider, *Dreaming the Biosphere* (Albuquerque: University of New Mexico Press, 2009).

8. A. Alling et al., "Biosphere 2 Test Module Experimentation Program," in *Biological Life Support Systems*, eds. M. Nelson and G. A. Soffen (Oracle, AZ: Synergetic Press, 1990).

9. M. Nelson and H. Bohn, "Soil-Based Biofiltration for Air Purification: Potentials for Environmental and Space Life Support Application," *Journal of Environmental Protection* 2, no. 8 (2011): doi: 10.4236/jep.2011.28125.

10. M. Nelson and B. C. Wolverton, "Plants + Soil/Wetland Microbes: Food Crop Systems That Also Clean Air and Water, *Advances in Space Research* 47 (2011): 582–590; C. Hodges and R. Frye, "Soil Bed Reactor Work of the Environmental

Research Lab of the University of Arizona in Support of the Biosphere 2 Project," in *Biological Life Support Systems—Commercial Opportunities*, eds. M. Nelson and G. A. Soffen (Tuscon, AZ: Synergetic Press, 1990): 33–40.

11. A. Abbott, "Scientists Bust Myth that Our Bodies Have More Bacteria than Human Cells," Nature.com, January 8, 2016, http://www.nature.com/news /scientists-bust-myth-that-our-bodies-have-more-bacteria-than-human-cells-1 .19136.

12. L. Sherwood, J. Willey, and C. J. Woolverton, *Prescott's Microbiology*, 9th ed. (New York: McGraw Hill, 2013).

13. Nelson and Wolverton, "Plants + Soil/Wetland Microbes."

14. A. Alling and M. Nelson, *Life under Glass: The Inside Story of Biosphere 2* (Oracle, AZ: Biosphere Press, 1993).

15. B. C. Wolverton, *Growing Fresh Air* (New York: Penguin, 1997).

16. "Seven Million Premature Deaths Annually Linked to Air Pollution," World Health Organization, March 25, 2014, http://www.who.int/mediacentre/news /releases/2014/air-pollution/en/.

17. J. Lelieveld et al., "The Contribution of Outdoor Air Pollution Sources to Premature Mortality on a Global Scale," *Nature* 525 (2015): 367–371.

18. Nelson and Bohn, "Soil-Based Biofiltration for Air Purification."

Chapter 5

1. W. H. Schlesinger, *Biogeochemistry: An Analysis of Global Change* (New York: Academic Publishers, 1991).

2. J. T. James et al., "Crew Health and Performance Improvements with Reduced Carbon Dioxide Levels and the Resource Impact to Accomplish Those Reductions," *American Institute of Aeronautics and Astronautics*, 2011, http://ntrs.nasa .gov/archive/nasa/casi.ntrs.nasa.gov/20100039645.pdf.

3. K. Kelly, *Out of Control: The New Biology of Machines, Social Systems and the Economic World* (New York: Addison Wesley Publishing Company, 1994).

4. J. Romm, "It's Taking Less CO_2 Than Expected to Cause Health Risks in Astronauts," ThinkProgress.org, October 28, 2015, https://thinkprogress.org /its-taking-less-co2-than-expected-to-cause-health-risks-in-astronauts.

5. B. G. Bugbee et al., "CO_2 Crop Growth Enhancement and Toxicity in Wheat and Rice," *Advances in Space Research* 14 (1994): 257–267.

6. "Ocean Acidification," National Geographic, April 27, 2017, http://ocean.national geographic.com/ocean/critical-issues-ocean-acidification/.

7. "Earth's CO_2 Home Page," CO_2 Earth, last modified June 2017, http://co2now .org/Current-CO2/CO2-Trend/.

8. "Forest Facts," American Forests, accessed April 17, 2017, https://www.american forests.org/discover-forests/tree-facts/.

9. R. Dahlman, G. Jacobs, and F. B. Metting, "What Is the Potential for Carbon Sequestration By the Terrestrial Biosphere?" National Energy Technology Laboratory, 2001, https://www.netl.doe.gov/publications/proceedings/01 /carbon_seq/5c0.pdf.

10. M. Lemonick, "Sources of Methane Emissions Still Uncertain: Study," Climate Central, January 30, 2014, http://www.climatecentral.org/news/sources-of -methane-emissions-still-uncertain-study-17010.

11. M. Nelson et al., "Atmospheric Dynamics and Bioregenerative Technologies in a Soil-Based Ecological Life Support System: Initial Results from Biosphere 2," *Advances in Space Research* 14, no. 11 (1994): 417–426.

12. Kelly, *Out of Control*, 159.

13. R. Schwartz, "Watch Chinese Air Pollution Work Its Way around the World in This Scary NASA Animation," Good.Is, February 2, 2015, http://magazine.good .is/articles/asian-air-pollution-spreading-around-the-world.

14. "Air and Water Pollution," Weather Explained, accessed April 12, 2017, http:// www.weatherexplained.com/Vol-1/Air-and-Water-Pollution.html.

15. "Forest Losses and Gains," Scribd, accessed September 11, 2017, https://www .scribd.com/document/316224552/Forest-Losses-and-Gains.

16. C. Lang, "20% of CO_2 Emissions from Deforestation? Make that 12," Redd, November 4, 2009, http://www.redd-monitor.org/2009/11/04/20-of-co2-emissions -from-deforestation-make-that-12/.

17. Stern Review, *The Economics of Climate Change*, UK Treasury report, 2006, http://webarchive.nationalarchives.gov.uk/20100407172811/http://www.hm -treasury.gov.uk/stern_review_report.htm.

18. "Regenerative Organic Culture and Climate Change," Rodale Institute, April 17, 2014, http://rodaleinstitute.org/regenerative-organic-agriculture-and-climate -change/.

19. U.S. Energy Information Administration, Frequently Asked Questions, published July 29, 2014, https://nnsa.energy.gov/sites/default/files/nnsa/08-14 -multiplefiles/DOE%202012.pdf.

20. B. Stevens, "Human Breathing and CO_2 (Carbon Dioxide): As Anthropogenic Causes of Climate Change and Global Warming," Barry on Energy, August 18, 2010, https://barryonenergy.wordpress.com/2010/08/18/human-breathing-and-co2 -carbon-dioxide-as-anthropogenic-causes-of-climate-change-and-global-warming/.

21. W. J. Broad, "Recycling Claim by Biosphere 2 Experiment Is Questioned," NYTimes.com, November 12, 1991, http://www.nytimes.com/1991/11/12/news /recycling-claim-by-biosphere-2-experiment-is-questioned.html?mcubz=0.

22. R. Monroe, "What Does 400 PPM Look Like?" Scripps Institution of Oceanography, December 3, 2013, https://scripps.ucsd.edu/programs/keelingcurve /2013/12/03/what-does-400-ppm-look-like/.

23. J. Bongaarts, "Human Population Growth and the Demographic Transition," *Philosophical Transactions of the Royal Society of London B: Biological Sciences* 364, no. 1,532 (October 27, 2009): 2,985–2,990, doi: 10.1098/rstb.2009.0137.

24. J. G. J. Olivier et al., "Trends in Global CO_2 Emissions 2013 Report," PBS Netherlands Environmental Assessment Agency, 2013, http://edgar.jrc.ec .europa.eu/news_docs/pbl-2013-trends-in-global-co2-emissions-2013-report -1148.pdf.

Chapter 6

1. L. Leigh et al., "An Introduction to the Intensive Agriculture Biome of Biosphere 2," in *Space Manufacturing 6: Nonterrestrial Resources, Biosciences and Space Engineering*, eds. B. Faughnan and G. Maryniak (Washington, DC: American Institute of Aeronautics and Astronautics, 1987): 76–81.

2. E. Glenn et al., "Sustainable Food Production for a Complete Diet," *HortScience* 25 (1990): 1,507–1,512.

3. R. Harwood, "There Is No Away," *Biosphere 2 Newsletter* 3, no. 3 (1993).

4. M. Nelson, S. Silverstone, and J. Poynter, "Biosphere 2 Agriculture: Test Bed for Intensive, Sustainable, Non-Polluting Farming Systems," *Outlook on Agriculture* 13, no. 3 (1993): 167–174.

5. S. Silverstone, *Eating In: From the Field to the Kitchen in Biosphere 2* (Oracle, AZ: Biosphere Press, 1993).

6. J. Poynter, *The Human Experiment: Two Years and Twenty Minutes inside Biosphere 2* (New York: Avalon Publishing Group, 2006).

7. R. Walford, *The 120-Year Diet* (New York: Simon and Schuster, 1987).

8. R. Reider, *Dreaming the Biosphere* (Albuquerque: University of New Mexico Press, 2009).

9. R. Walford, S. B. Harris, and M. W. Gunion, "Calorically Restricted Low-Fat Nutrient-Dense Diet in Biosphere 2 Significantly Lowers Blood Glucose, Total Leukocyte Count, Cholesterol, and Blood Pressure in Humans," *Proceedings, National Academy of Sciences* 89, no. 23 (1992): 11,533–11,537.

10. Silverstone, *Eating In*.

11. L. Schumm, "America's Patriotic Victory Gardens," History.com, May 29, 2014, http://www.history.com/news/hungry-history/americas-patriotic-victory -gardens.

12. S. Silverstone and M. Nelson, "Food Production and Nutrition in Biosphere 2: Results from the First Mission, September 1991 to September 1993," *Advances in Space Research* 18, no. 4/5 (1996): 49–61.

13. R. Harwood, "There Is No Away."

14. Silverstone, *Eating In*.

15. B. D. V. Marino et al., "The Agricultural Biome of Biosphere 2: Structure, Composition and Function," *Ecological Engineering* 13, no. 1–4 (1999): 199–234.

16. Poynter, *The Human Experiment.*

17. Ibid.

18. Silverstone, *Eating In.*

19. Ibid.

20. Ibid.

21. "Food Miles: How Far Your Food Travels Has Serious Consequences for Your Health and the Climate," Natural Resources Defense Council, November 2007, https://food-hub.org/files/resources/Food%20Miles.pdf.

22. "Potato Facts and Figures," International Potato Center, accessed March 10, 2017, http://cipotato.org/potato/facts/.

23. S. Critchfield, "We Used to Have 307 Kinds of Corn. Guess How Many Are Left?" Upworthy.com, April 18, 2012, http://www.upworthy.com/we-used-to -have-307-kinds-of-corn-guess-how-many-are-left.

24. "Diminished Crop Diversity," Oregon State University, last modified November 18, 2011, http://people.oregonstate.edu/~muirp/cropdiv.htm.

25. "Biodiversity to Nurture People," FAO Corporate Document Repository, accessed March 13, 2017, http://www.fao.org/docrep/004/v1430e/V1430E04.htm.

26. C. K. Khoury et al., *Measuring the State of Conservation of Crop Diversity: A Baseline for Marking Progress Toward Biodiversity Conservation and Sustainable Development Goals*, policy brief, CGIAR, 2016, https://ccafs.cgiar.org /publications/measuring-state-conservation-crop-diversity-baseline-marking -progress-toward#.WK8xaFXyuUl.

27. D. Pimental, "Soil Erosion: A Food and Environmental Threat," *Environment, Development and Sustainability* 8 (2006): 109–117.

28. R. Walford et al., "Biospheric Medicine as Viewed from the 2-Year Closure of Biosphere 2," *Aviation, Space and Environmental Medicine* 67, no. 7 (1996): 609–617.

29. R. Goodland, "Livestock and Climate Change," *WorldWatch*, November/ December 2009, https://www.worldwatch.org/files/pdf/Livestock%20and%20 Climate%20Change.pdf.

30. M. M. Mekonnen and A. Y. Hoekstra, "The Green, Blue and Grey Water Footprint of Crops and Derived Crop Products," UNESCO-IHE Institute for Water Education, 2010, http://waterfootprint.org/media/downloads/Report47 -WaterFootprintCrops-Vol1.pdf.

31. "The Number of Vegetarians in the World," Raw Food Health, accessed April 20, 2017, http://www.raw-food-health.net/NumberOfVegetarians.html.

32. J. L. P., "Meat and Greens," *The Economist*, December 13, 2013, http://www .economist.com/blogs/feastandfamine/2013/12/livestock.

33. R. Nuwer, "Raising Beef Uses Ten Times More Resources Than Poultry, Dairy, Eggs or Pork," Smithsonian.com, July 21, 2014, http://www.smithsonian mag.com/science-nature/beef-uses-ten-times-more-resources-poultry-dairy -eggs-pork-180952103/?no-ist.

34. S. Kumar, "The Forgotten 1 Billion," Worldwatch Institute, December 21, 2011, http://www.worldwatch.org/forgotten-1-billion.

35. M. Nelson, "Biosphere 2 Journal," 1991–1993.

36. S. Silverstone et al., "Soil in the Agricultural Area of Biosphere 2 (1991–1993)," *Ecological Engineering* 13 (1999): 179–188.

Chapter 7

1. M. Nelson and B. C. Wolverton, "Plants + Soil/Wetland Microbes: Food Crop Systems That Also Clean Air and Water," *Advances in Space Research* 47 (2011): 582–590.

2. M. Nelson et al., "Bioregenerative Recycle of Wastewater in Biosphere 2 Using a Created Wetland: Two Year Results," *Ecological Engineering* 13 (1999): 189–197.

3. M. Nelson, *The Wastewater Gardener: Preserving the Planet One Flush At a Time* (Santa Fe, NM: Synergetic Press, 2014).

4. *Chemicals in the Environment: Chlorine* (CAS NO. 7782–50–5) (1994) Office of Pollution Prevention and Toxics, U.S. Environmental Protection Agency August 1994, http://www.epa.gov/chemfact/f_chlori.txt; *Primer for Municipal Wastewater Treatment Systems*, Environmental Protection Agency, document no. EPA 832-R-04–001 (Washington, DC: 2004).

5. D. Spector, "Our Planet Is Exploding with Ocean Dead Zones," Business Insider, June 26, 2013, http://www.businessinsider.com/map-of-worldwide -marine-dead-zones-2013-6.

6. "Diarrhoeal Disease," World Health Organization, last modified May 2017, http://www.who.int/mediacentre/factsheets/fs330/en/.

7. World Water Day report, World Health Organization, accessed August 29, 2017, http://www.who.int/water_sanitation_health/takingcharge.html.

8. F. H. King, *Farmers of Forty Centuries* (Emmaus, PA: Rodale Press, 2011).

9. J. Jenkins, *The Humanure Handbook: A Guide to Composting Human Manure*, 2nd ed. (Grove City, PA: Joseph Jenkins, Inc., 1999).

10. "Backyard Composting," Seattle Public Utilities, accessed April 16, 2017, http:// www.seattle.gov/util/EnvironmentConservation/MyLawnGarden/Compost Soil/Composting/index.htm.

11. Nelson, *The Wastewater Gardener.*

12. K. Chung and M. White, "Greywater Reuse," Cornell University courses, accessed April 17, 2017. https://courses.cit.cornell.edu/crp384/2009reports/White&Chung_Gray%20Water%20Reuse.pdf.

13. "Water Facts and Figures," International Fund for Agricultural Development, accessed April 17, 2017, http://www.ifad.org/english/water/key.htm.

14. J. Diamond, "The Erosion of Civilization: The Fertile Crescent's Fall Holds a Message for Today's Troubled Spots," Los Angeles Times, June 15, 2003, http://articles.latimes.com/2003/jun/15/opinion/op-diamond15.

15. D. R. Montgomery, Dirt: The Erosion of Civilizations (Berkeley, CA: University of California Press, 2007), 295.

16. J. Diamond, Collapse: How Societies Choose to Fail or Succeed (New York: Penguin, 2005).

17. P. Rengasamy, "Salinity in the Landscape: A Growing Problem in Australia," Geotimes: Earth, Energy and Environmental News, March 2008, http://www.geotimes.org/mar08/article.html?id=feature_salinity.html.

18. R. Munns and J. Passioura, "The Dirt on Our Soils," Nova: Australian Academy of Science, accessed April 19, 2017, http://www.nova.org.au/earth-environment/dirt-our-soils.

19. "Salinity in the Central Valley: A Critical Problem," Water Education Foundation, accessed April 21, 2017, http://www.watereducation.org/post/salinity-central-valley-critical-problem.

20. A. Bot and J. Benites, "Natural Factors Influencing the Amount of Organic Matter," in The Importance of Soil Organic Matter, FAO Corporate Document Repository, 2005, http://www.fao.org/docrep/009/a0100e/a0100e06.htm.

21. I. P. Abrol, J. S. P. Yadav and F. I. Massoud, "Salt-Affected Soils and Their Management," FAO Soils Bulletin 39 (1988): http://www.fao.org/docrep/x5871e/x5871e00.htm.

22. "Effects of Acid Rain," United States Environmental Protection Agency, accessed August 29, 2017, https://www.epa.gov/acidrain/effects-acid-rain.

23. "Water Topics," Environmental Protection Agency, accessed April 22, 2017, http://water.epa.gov/action/weatherchannel/stormwater.cfm.

Chapter 8

1. U.S. Forest Facts and Historical Trends, U.S. Department of Agriculture, September 2001, https://www.fia.fs.fed.us/library/brochures/docs/2000/Forest-FactsMetric.pdf; J. O. Kaplan, K. M. Krumhardt, and N. Zimmerman, The Prehistoric and Preindustrial Deforestation of Europe," Quarternary Science Reviews 28 (2009): 3,016–3,034, doi:10.1016/j.quascirev.2009.09.028.; C. J. A.

Bradshaw, "Little Left to Lose: Deforestation and Forest Degradation in Australia Since European Colonization," *Journal of Plant Ecology* 5, no. 1 (March 2012): 109–120, doi: 10.1093/jpe/rtr038.

2. European Commission, Joint Research Center, "Urbanization: 95% of the World's Population Lives on 10% of the Land," December 19, 2008, https://www.sciencedaily.com/releases/2008/12/081217192745.htm; J. Owen, "Farming Claims Almost Half Earth's Land, New Maps Show," *National Geographic News*, December 9, 2005, http://news.nationalgeographic.com/news/2005/12/1209_051209_crops_map.html.

3. R. Nuwer, "Raising Beef Uses Ten Times More Resources Than Poultry, Dairy, Eggs or Pork," Smithsonian.com, July 21, 2014, http://www.smithsonianmag.com/science-nature/beef-uses-ten-times-more-resources-poultry-dairy-eggs-pork-180952103/?no-ist.

4. "The Tropical Rainforest," University of Michigan, November 16, 2016, http://www.globalchange.umich.edu/globalchange1/current/lectures/kling/rainforest/rainforest.html.

5. "Keystone Species," *National Geographic*, accessed August 26, 2017, http://education.nationalgeographic.com/encyclopedia/keystone-species/.

6. M. Grunwald, *The Swamp: The Everglades, Florida, and the Politics of Paradise* (New York: Simon & Schuster, 2006).

7. University of California, Santa Barbara, "Earth in Midst of Sixth Mass Extinction: 50% of All Species Disappearing," ScienceDaily.com, October 21, 2008, https://www.sciencedaily.com/releases/2008/10/081020171454.htm.

8. "The Extinction Crisis," Center for Biological Diversity, accessed April 26, 2017, http://www.biologicaldiversity.org/programs/biodiversity/elements_of_biodiversity/extinction_crisis/.

9. J. Petersen et al., "The Making of Biosphere 2," *Ecological Restoration* 10, no. 2 (1992): 158–168.

10. M. E. Kentula, "Wetland Restoration and Creation," National Water Summary on Wetland Resources, *United States Geological Survey Water Supply Paper 2425*, last modified January 29, 2002, https://water.usgs.gov/nwsum/WSP2425/restoration.html.

11. J. P. Allen, *Biosphere 2: The Human Experiment* (New York: Penguin, 1991).

12. G. C. Daily et al., "Ecosystem Services: Benefits Supplied to Human Societies by Natural Ecosystems," *Issues in Ecology*, no. 2 (Spring 1997).

13. D. J. McGee et al., Bacterial Genetic Exchange, 2001. eLS. http://onlinelibrary.wiley.com/doi/10.1038/npg.els.0001416/full.

14. E. O. Wilson, "The Little Things That Run the World: The Importance and Conservation of Invertebrates," *Conservation Biology* 1 (1987): 344–346.

15. Allen, *Biosphere 2: The Human Experiment*.

16. K. Kelly, "Biosphere 2 at One," *Whole Earth Review*, no. 77 (winter 1992): 90–105.

17. Ibid.

Chapter 9

1. R. Sears, "Tepui," World Wildlife Fund, accessed May 2, 2017, http://www
   .worldwildlife.org/ecoregions/nt0169.

2. "Amazon Floodplain Forests," World Wildlife Fund, accessed April 30, 2017,
   http://wwf.panda.org/what_we_do/where_we_work/amazon/about_the
   _amazon/ecosystems_amazon/floodplain_forests/.

3. H. T. Odum, "Scales of Ecological Engineering," *Ecological Engineering* 6
   (1996): 7–19.

4. L. Leigh, et al., "Tropical Rainforest Biome of Biosphere 2: Structure, Composition and Results of the First 2 Years of Operation," *Ecological Engineering* 13
   (1999): 65–93.

5. L. Leigh, "The Basis for Rainforest Diversity in Biosphere 2," (PhD diss., University of Florida, Gainesville, 1999), 364.

6. J. K. Wetterer et al., "Ecological Dominance by *Paratrechina longicornis* (Hymenoptera: Formicidae), an Invasive Tramp Ant, in Biosphere 2," *Florida Entomologist* 82 (1999): 381–388.

7. A. Alling and M. Nelson, *Life Under Glass: The Inside Story of Biosphere 2* (Oracle, AZ: Biosphere Press, 1993).

8. Leigh, "The Basis for Rainforest Diversity."

9. W. Laurance et al., "The Fate of Amazonian Forest Fragments: A 32-Year
   Investigation," *Biological Conservation* 144 (2011): 56–67, doi:10.1016/j.biocon
   .2010.09.021.

10. "Chitwan National Park," UNESCO, accessed May 4, 2017, http://whc.unesco
    .org/en/list/284.

11. B. M. Boom, "Amazonian Indians and the Forest Environment," *Nature* 314
    (1984): 324.

12. R. Butler, "The Effect of Area on Rainforest Species Richness," Mongabay, last
    modified July 31, 2012, http://rainforests.mongabay.com/0303a.htm.

13. M. R. Hands, A. F. Harrison, and T. P. Bayliss-Smith, "Phosphorus Dynamics
    in Slash-and-Burn and Alley-Cropping Systems of the Humid Tropics," in
    Phosphorus in the Global Environment, ed. H. Tiessen (New York: John Wiley,
    1995); R. Sitler, "Providing an Alternative to Slash-and-Burn Agriculture,"
    October 21, 2013, http://www.resilience.org/stories/2013-10-21/providing-an
    -alternative-to-slash-and-burn-agriculture.

14. L. Taylor, "Rainforest Facts," last modified December 21, 2012, http://www.rain
    -tree.com/facts.htm#.VUF_GfldXVQ.

15. "Amazon Rainforest," BluePlanetBiomes.org, 2003, http://www.blueplanet biomes.org/amazon.htm.

16. L. Taylor, "Saving the Rainforest: A Complex Problem and a Simple Solution," The Raintree Group, Inc., accessed June 15, 2017, http://csc.columbusstate.edu /summers/Outreach/RainSticks/fRainforestFacts.htm.

17. Taylor, "Rainforest Facts."

18. M. Nelson et al., "Enriched Secondary Subtropical Forest through Line-Planting for Sustainable Timber Production in Puerto Rico," *Bois et Forêts des Tropiques* 309, no. 3 (2011): 51–63; M. Nelson et al., "The Impact of Hardwood Line-Planting on Tree and Amphibian Biodiversity in a Secondary Wet Tropical Forest, Southeast Puerto Rico," *Journal of Sustainable Forestry* 29, no. 5 (2010): 503–516.

19. Rainforest Alliance, "Halting Deforestation and Achieving Sustainability," accessed September 12, 2017, http://www.rainforest-alliance.org/sites/default /files/2016-08/Deforestation-and-Sustainability-RA-position-paper-2015-04 -13_1_0.pdf.

20. A. Molnar et al., "Community-Based Forest Management: The Extent and Potential Scope of Community and Smallholder Forest Management and Enterprise" (Washington, DC: Rights and Resources Initiative, 2011), https://commdev.org/userfiles/Community%20based%20forest%20 management.pdf.

Chapter 10

1. C. C. Gaither and A. E. Cavazos-Gaither, *Gaither's Dictionary of Scientific Quotations* (Springer: New York, 2008), 1,636.

2. "Coral Reefs: Tropical Corals," World Wildlife Fund, accessed May 15, 2017, http://wwf.panda.org/about_our_earth/blue_planet/coasts/coral_reefs /tropical_corals/.

3. "In What Types of Waters Do Corals Live?" National Ocean Service, accessed August 29, 2017, https://oceanservice.noaa.gov/facts/coralwaters.html.

4. W. Adey and K. Loveland, *Dynamic Aquaria* (San Diego: Academic Press, 1991), 463.

5. R. Scarborough, "The Geologic Origins of the Sonoran Desert," Arizona-Sonora Desert Museum, accessed May 15, 2017, https://www.desertmuseum .org/books/nhsd_geologic_origin.php.

6. R. Reider, *Dreaming the Biosphere* (Albuquerque: University of New Mexico Press, 2009).

7. P. R. Escobal, *Aquatic Systems Engineering: Devices and How They Function* (Dimension Engineering Press, 2000); R. Holmes-Farley, "What is Skimming?",

Reefkeeping, August 2006, http://www.reefkeeping.com/issues/2006-08/rhf/index.php.

8. A. Alling and M. Nelson, *Life Under Glass: The Inside Story of Biosphere 2* (Oracle, AZ: Biosphere Press, 1993).

9. Ibid.

10. "Results of Biosphere 2," Biospherics.org, accessed April 25, 2017, http://www.biospherics.org/biosphere2/results/.

11. Reider, *Dreaming the Biosphere*.

12. D. M. Spoon, C. J. Hogan, and G. B. Chapman, "Ultrastructure of a Primitive, Multinucleate, Marine, Cyanobacteriophagous Ameba (*Euhyperamoeba biospherica*, n.sp.) and Its Possible Significance in the Evolution of Lower Eukaryotes," *Invertebrate Biology* 114, no. 3 (1995): 189–201.

13. D. M. Spoon, and A. Alling, "Preclosure Survey of Aquatic Microbiota of Biosphere 2," paper presented at the Third East Coast Conference on Protozoa, Mount Vernon College, VA, May 21–22, 1991; C. Luckett et al., "Coral Reef Mesocosms and Microcosms: Successes, Problems, and the Future of Laboratory Models, *Ecological Engineering* 6, no. 1–3 (1996): 57–72.

14. Reider, *Dreaming the Biosphere*.

15. "Latest Review of Science Reveals Ocean in Critical State," IUCN, October 3, 2013, https://www.iucn.org/content/latest-review-science-reveals-ocean-critical-state.

16. K. Olds, P. Dustan, and A. Alling, "Eight Years of Coral Reef Data from Melanesia and Southeast Asia: Vitality, Percent Cover, Reef Cover and Effects of Earthquake," poster prepared for the International Coral Reef Symposium, Florida, 2008, http://www.pcrf.org/pdf/ICRSKatie08.pdf.

17. P. Dustan, "Developing Methods for Assessing Coral Reef Vitality: A Tale of Two Scales," in *Proceedings of the Colloquium on Global Aspects of Coral Reefs: Health, Hazards, and History*, ed. R. N. Ginsburg (Miami, FL: University of Miami, 1994), 38–44.

18. "Planetary Coral Reef Expedition (Research)," Biosphere Foundation, accessed September 2, 2017, https://biospherefoundation.org/project/pcrf-planetary-coral-reef-foundation/.

19. J. Freund, "Coral Reef: Threats," World Wildlife Fund, accessed September 2, 2017, http://wwf.panda.org/about_our_earth/blue_planet/coasts/coral_reefs/coral_threats.

20. "Oceans and Coasts," The Economics of Ecosystems & Biodiversity, accessed September 2, 2017, http://www.teebweb.org/areas-of-work/biome-studies/teeb-for-oceans-and-coasts/.

21. "The Water Cycle: The Oceans," U.S. Geological Survey, last modified December 2, 2016, https://water.usgs.gov/edu/watercycleoceans.html.

22. J. Sultan, "Sea-News: The Demise of Phytoplankton, Earth's Ultimate Producer," Mission Blue, December 31, 2010, http://mission-blue.org/2010/12/sea-news -the-demise-of-phytoplankton-earths-ultimate-producer/.

23. M. Ateweberhana et al., "Climate Change Impacts on Coral Reefs: Synergies with Local Effects, Possibilities for Acclimation, and Management Implications," *Marine Pollution Bulletin* 74, no. 2 (September 30, 2013): 526–539.

24. A. Alling et al., "Catastrophic Coral Mortality in the Remote Central Pacific Ocean: Kiribati Phoenix Islands, *Atoll Research Bulletin* no. 551 (National Museum of Natural History, Smithsonian Institution, December 2007), http:// www.pcrf.org/pdf/CantonPaper08.pdf.

25. "Overfishing and Destructive Fishing Threats," Reef Resilience, accessed August 30, 2017, http://www.reefresilience.org/coral-reefs/stressors/local -stressors/overfishing-and-destructive-fishing-threats/.

26. The Ocean Portal Team, "Deep Sea Corals," Ocean Portal, accessed June 1, 2017, http://ocean.si.edu/deep-sea-corals.

27. "Threats: Bycatch," World Wildlife Fund, accessed April 20, 2017, https://www .worldwildlife.org/threats/bycatch.

28. Ibid.; S. Dolman, "End Bycatch: Stop Deaths in Fishing Gear," Whale and Dolphin Conservation, accessed April 25, 2017, us.whales.org/issues/fishing.

29. "Analysis of U.S. MPAs," Marine Protected Areas, last modified July 31, 2017, http://marineprotectedareas.noaa.gov/dataanalysis/analysisus/.

30. P. Goriup, *Protected Areas Programme* 17, no. 2 (2008): https://cmsdata.iucn.org /downloads/parks_17_2_web.pdf.

31. "What Is a Marine Protected Area?" National Ocean Service, accessed April 25, 2017, http://oceanservice.noaa.gov/facts/mpa.html.

32. D. d'A. Laffoley, ed., *Towards Networks of Marine Protected Areas* (Gland, Switzerland: IUCN WCPA, 2008), 28.

33. M. Christie et al., "Larval Connectivity in an Effective Network of Marine Protected Areas, *PLoS ONE* 5, no. 12 (2010): doi:10.1371/journal.pone.0015715; Oregon State University, "Drifting Fish Larvae Allow Marine Reserves to Rebuild Fisheries," *Science Daily*, December 26, 2010, https://www.sciencedaily.com /releases/2010/12/101222173105.htm.

34. D. Montalvo, "Hawaii's Bleaching Problem: How Warming Waters Threaten Coral," CNBC, July 18, 2015, http://www.cnbc.com/2015/07/18/hawaiis-coral -threatened-by-bleaching.html.

Chapter 11

1. M. Finn, "Mangrove Mesocosm of Biosphere 2: Design, Establishment and Preliminary Results," *Ecological Engineering* 6 (1996): 21–56.

2. W. Adey and K. Loveland, *Dynamic Aquaria* (San Diego: Academic Press, 1991), 463.

3. A. Alling and M. Nelson, *Life under Glass: The Inside Story of Biosphere 2* (Oracle, AZ: Biosphere Press, 1993).

4. M. Finn, *Comparison of Mangrove Forest Structure and Function in a Mesocosm and Florida* (PhD diss., Georgetown University, 1996).

5. M. Nelson, "Litter Fall and Decomposition Rates in Biosphere 2 Terrestrial Biomes," *Ecological Engineering* 13 (1999): 135–145.

6. Finn, *Comparison of Mangrove Forest Structure.*

7. "What's a Mangrove? And How Does It Work?" American Museum of Natural History, accessed September 5, 2017, https://www.amnh.org/explore /science-bulletins/bio/documentaries/mangroves-the-roots-of-the-sea/what-s -a-mangrove-and-how-does-it-work/.

8. A. B. Rath, "Mangrove Importance," World Wildlife Fund, accessed May 12, 2017, http://wwf.panda.org/about_our_earth/blue_planet/coasts/mangroves /mangrove_importance/.

9. S. R. Massel, "Tides and Waves in Mangrove Forests, in *Fluid Mechanics for Marine Ecologists* (Berlin: Springer-Verlag, 1999), 418–425.

10. M. Spalding et al., *Mangroves for Coastal Defense*, Wetlands International and the Nature Conservatory, 2014, http://www.mangrovesforthe future.org/assets/Repository/Documents/WI-TNC-mangroves-for-coastal -defence.pdf.

11. International Union for the Conservation of Nature, *Early Observations of Tsunami Effects on Mangroves and Coastal Forests. Statement from the IUCN Forest Conservation Programme*, January 7, 2005, accessed March 17, 2017, http://www .iucn.org/info_and_news/press.pdf.

12. "Sundarbans Mangroves," World Wildlife Fund, accessed August 30, 2017, http:// wwf.panda.org/about_our_earth/ecoregions/sundarbans_mangroves.cfm.

13. S. Wells, C. Ravilious, and E. Corcoran, "In the Front Line: Shoreline Protection and Other Ecosystem Services from Mangroves and Coral Reefs," UNEP-WCMC (2006): 33.

14. I. Valiela, J. L. Bowen, and J. K. York, "Mangrove Forests: One of the World's Threatened Major Tropical Environments," *Bioscience* 51, no. 10 (2001): 807–815.

15. J. Hutchinson et al., "Predicting Global Patterns in Mangrove Forest Biomass," *Letters in Conservation* 7, no. 3 (2014): 233–240.

16. M. D. Spalding, "The Global Distribution and Status of Mangrove Ecosystems, Mangrove Edition," *International Newsletter of Coastal Management* (Intercoast Network) Special Edition #1 (Narragansett: Coastal Resources Center, University of Rhode Island, 1997), 20–21.

17. Wells, Ravilious, and Corcoran, "In the Front Line," 33.

18. M. Finn, P. Kangas, and W. Adey, "Mangrove Ecosystem Development in Biosphere 2," *Ecological Engineering* 13 (1999): 173–178.

Chapter 12

1. C. Q. Choi, "Savanna, Not Forest, Was Human Ancestors' Proving Ground," LiveScience, August 3, 2011, http://www.livescience.com/15377-savannas -human-ancestors-evolution.html.

2. J. H. Falk and J. D. Balling, "Evolutionary Influence on Human Landscape Preference," *Environment and Behavior* 42, no. 4 (2010): 479–493, doi: 10.1177/ 0013916509341244.

3. T. L. McKnight and D. Hess, "Climate Zones and Types," in *Physical Geography: A Landscape Appreciation* (Upper Saddle River, NJ: Prentice Hall, 2000).

4. A. Alling and M. Nelson, *Life under Glass: The Inside Story of Biosphere 2* (Oracle, AZ: Biosphere Press, 1993).

5. "Time Line of the American Bison," U.S. Fish and Wildlife Service, accessed September 5, 2017, http://www.fws.gov/bisonrange/timeline.htm.

6. G. L. Stebbins, "Coevolution of Grasses and Herbivores," *Annals of the Missouri Botanical Garden* 68, no. 1 (1981): 75–86.

7. M. Nelson, "Synergetic Management of the Savannas," in *Ecology and Management of the World's Savannas*, eds. J. C. Tothill and J. J. Mott (Brisbane: University of Queensland Press, 1985).

8. "Climate of the Tropical Savannas," Savanna Explorer, accessed May 10, 2017, http://www.savanna.org.au/all/climate.html.

9. J. C. Tothill and J. J. Mott, eds., *Ecology and Management of the World's Savannas*.

10. "Tropical Savannas," Biomes of the World, accessed May 10, 2017, https://php .radford.edu/~swoodwar/biomes/?page_id=105.

11. B. Gammage, *The Biggest Estate on Earth: How Aborigines Made Australia* (Sydney, Australia: Allen & Unwin, 2011).

12. N. H. Speck et al., "No. 9 General Report on Lands of the West Kimberley Area, W.A. CSIRO Land Surveys," *Land Research Surveys* 1 (2010): 1–228.

13. E. Huntington, *A System of Modern Geography* (1834).

14. A. Kirby, "Climate Renews Famine Risk to Africa's Sahel," Climate News Network, October 20, 2014, http://www.climatenewsnetwork.net/climate-renews -famine-risk-to-africas-sahel/.

15. A. Spanne, "Industrial Farming Plows Up Brazil's 'Underground Forest,'" Climate Central, November 15, 2014, http://www.climatecentral.org/news /industrial-farming-brazil-cerrado-18332.

16. J. M. C. Da Silva and J. M. Bates, "Biogeographic Patterns and Conservation in the South American Cerrado: A Tropical Savanna Hotspot," *BioScience* 52, no. 3 (March 2002): 225–234.

17. J. Pickrell, "Trophy Hunting Can Help African Conservation, Study Says," *National Geographic News*, March 15, 2007, http://news.nationalgeographic .com/news/2007/03/070315-hunting-africa.html.

18. "The State of the Lion," Panthera.org, accessed May 12, 2017, http://www .panthera.org/node/8.

19. "Tiger," World Wildlife Fund, accessed May 15, 2017, http://www.worldwildlife .org/species/tiger.

20. A. Gabor, "Inside Polyface Farm, Mecca of Sustainable Agriculture," *The Atlantic*, July 5, 2011, https://www.theatlantic.com/health/archive/2011/07/inside -polyface-farm-mecca-of-sustainable-agriculture/242493/.

Chapter 13

1. F. Norte, "Fog Desert," in M. A. Mares, *Encyclopedia of Deserts* (Norman, OK: University of Oklahoma Press, 1999).

2. A. Valero, J. Schipper, and T. Allnutt, "Southern North America: Baja California Peninsula in Mexico," World Wildlife Fund, accessed March 17, 2017, http:// www.worldwildlife.org/ecoregions/na1301.

3. A. Alling and M. Nelson, *Life under Glass: The Inside Story of Biosphere 2* (Oracle, AZ: Biosphere Press, 1993).

4. K. Kelly, "Biosphere 2 at One," *Whole Earth Review* 77 (winter 1992): 90–105.

5. J. K. Wetterer et al., "Ecological Dominance by *Paratrechina longicornis* (Hymenoptera: Formicidae), an Invasive Tramp Ant, in Biosphere 2," *Florida Entomologist* 82 (1999): 381–388.

6. J. E. Cohen and D. Tilman, "Biosphere 2 and Biodiversity: The Lessons So Far," *Science* 274, no. 5,290 (1996): 1,150–1,151.

7. Cohen and Tilman, "Biosphere 2 and Biodiversity."

8. W. Mitsch, "Preface," *Ecology Engineering* 13 (1999): 1–2.

9. Cohen and Tilman, "Biosphere 2 and Biodiversity."

10. B. Bryson, *A Short History of Nearly Everything* (New York: Broadway Books, 2003), 350–370.

11. B. P. Stearns and S. C. Stearns, *Watching, from the Edge of Extinction* (New Haven, CT: Yale University Press, 2000).

12. "The Current Mass Extinction," PBS, accessed March 8, 2017, http://www.pbs .org/wgbh/evolution/library/03/2/l_032_04.html.

Chapter 14

1. W. Broecker, "Et Tu, O_2?" 21st C at Columbia University, accessed April 19, 2017, http://www.columbia.edu/cu/21stC/issue-2.1/broecker.htm.

2. R. L. Walford et al., "Biospheric Medicine as Viewed from the Two-Year First Closure of Biosphere 2," *Aviation, Space, and Environmental Medicine* 67, no. 7 (1996): 609–617.

3. S. Lonstaff, "The Unexamined Life Is Not Worth Living," New Philosopher, June 2, 2013, http://www.newphilosopher.com/articles/being-fully-human/.

4. H. A. Torbert and H. B. Johnson, "Soil of the Intensive Agriculture Biome of Biosphere 2," *Journal of Soil and Water Conservation* 56, no. 1 (2001): 4–11.

5. C. Weyer et al., "Energy Metabolism after 2 Years of Energy Restriction: The Biosphere 2 Experiment," *The American Journal of Clinical Nutrition* 72, no. 4 (2000): 946–953.

6. L. Leigh, "Linda's Journal—Oxygen," *Biosphere 2 Newsletter* 3, no. 1 (1993).

7. R. L. Walford and S. R. Spindler, "The Response to Calorie Restriction in Mammals Shows Features Also Common to Hibernation, a Cross-Adaptation Hypothesis," *Journal of Gerontology, Biological Sciences* 52A, no. 4 (1997).

8. D. Herring and R. Kannenberg, "The Mystery of the Missing Carbon," NASA Earth Observatory, 1999, http://earthobservatory.nasa.gov/Features /BOREASCarbon/.

9. "Scientists Close In On Missing Carbon Sink," NCAR News Release, June 21, 2007, http://www.ucar.edu/news/releases/2007/carbonsink.shtml.

10. H. T. Odum, "Scales of Ecological Engineering," *Ecological Engineering* 6 (1996): 7–19.

11. V. C. Engel and H. T. Odum, "Simulations of Community Metabolism and Atmospheric Carbon Dioxide and Oxygen Concentrations in Biosphere 2," *Ecological Engineering* 13 (1999): 107–134.

Chapter 15

1. M. Roach, *Packing for Mars: The Curious Science of Life in the Void* (New York: W. W. Norton & Co., 2010).

2. R. E. Byrd, *Alone*, (New York: Putnam Books, 1938), quoted in A. F. Barabasz, "A Review of Antarctic Behavioral Research," in *From Antarctica to Outer Space: Life in Isolation and Confinement*, eds. A. A. Harrison, Y. A. Clearwater, and C. P. McKay (New York: Springer-Verlag, 1991), 21–30.

3. "Biosphere 2 Plus 15," Living on Earth, March 17, 2006, http://www.loe.org /shows/segments.html?programID=06-P13-00011&segmentID=3.

4. W. R. Bion, *Experiences in Groups: And Other Papers* (New York: Routledge, 1968).

5. H. T. Odum, "Scales of Ecological Engineering," *Ecological Engineering* 6 (1996): 7–19.

6.   C. Fehrman, "The Incredible Shrinking Sound Bite," Boston.com, January 2, 2011, http://www.boston.com/bostonglobe/ideas/articles/2011/01/02/the_incredible_shrinking_sound_bite/.

7.   "Biospherians Mum on Birds and Bees," *Arizona Republic*, September 25, 1993, https://www.newspapers.com/newspage/123014124/.

8.   J. Poynter, *The Human Experiment: Two Years and Twenty Minutes inside Biosphere 2* (New York: Avalon Publishing Group, 2006).

9.   J. Allen and M. Nelson, "Biospherics and Biosphere 2, Mission One (1991–1993)," *Ecological Engineering* 13 (1999): 15–29.

10.   M. Nelson, K. Gray, and J. P. Allen, "Group Dynamics as a Critical Component of Space Exploration and Terrestrial CELSS with Humans: Conceptual Theory and Insights from Mission One of Biosphere 2," *Life Sciences in Space Research* 6 (2015): 79–86.

11.   "Biosphere 2 Plus 15."

12.   R. L. Walford, "Biosphere 2 as Voyage of Discovery: The Serendipity from Inside," *Bioscience* 52, no. 3 (2002): 259–263.

13.   N. Kanas and D. Manzey, *Space Psychology and Psychiatry* (El Segundo, CA: Microcosm Press and Springer, 2003).

14.   Roach, *Packing for Mars.*

15.   R. Reider, *Dreaming the Biosphere* (Albuquerque: University of New Mexico Press, 2009).

16.   Kanas and Manzey, *Space Psychology and Psychiatry.*

17.   Poynter, *The Human Experiment.*

18.   Reider, *Dreaming the Biosphere.*

19.   Ibid.

20.   Ibid.

21.   Odum, "Scales of Ecological Engineering."

22.   Poynter, *The Human Experiment.*

23.   Walford, "Biosphere 2 as Voyage of Discovery."

24.   J. P. Allen, "People Challenges in Biospheric Systems for Long-Term Habitation in Remote Areas, Space Stations, Moon, and Mars Expeditions," *Life Support and Biosphere Science* 8 (2002): 67–70.

25.   A. Alling et al., "Human Factor Observations of the Biosphere 2, 1991–1993, Closed Life Support Human Experiment and Its Application to a Long-Term Manned Mission to Mars," *Life Support and Biosphere Science* 8 (2002): 71–82.

26.   O. Gazenko, personal communication with K. Gray, September 28, 1993.

27.   D. Oliver, "Psychological Effects of Isolation and Confinement of a Winter-Over Group at McMurdo Station Antarctica," in *From Antarctica to Outer Space: Life in Isolation and Confinement*, eds. A. A. Harrison, Y. A. Clearwater, and C. P. McKay (New York: Springer-Verlag, 1991), 217–227.

28.   Poynter, *The Human Experiment.*

29. R. B. Bechtel, T. MacCallum, and J. Poynter, "Environmental Psychology and Biosphere 2," in: *Handbook of Japan-United States Environment-Behavior Research* (New York: Springer, 1997), 235–244.

30. M. Nelson, private Biosphere 2 journal, 1991–1993, September 24, 1993.

31. Alling et al., "Human Factor Observations of the Biosphere 2"; A. Alling and M. Nelson, *Life under Glass: The Inside Story of Biosphere 2* (Oracle, AZ: Biosphere Press, 1993).

32. Alling et al., "Human Factor Observations of the Biosphere 2.".

33. Nelson, Gray, and Allen, "Group Dynamics as a Critical Component of Space Exploration."

34. Reider, *Dreaming the Biosphere.*

35. G. Albrecht et al., "Solastalgia: The Distress Caused by Environmental Change," *Australas Psychiatry* 15, no. 1 (2007): 95–98.

36. N. Klein, *This Changes Everything: Capitalism vs. the Climate* (New York: Simon and Shuster, 2014).

37. E. O. Wilson, *Biophilia* (Cambridge, MA: Harvard University Press, 1984).

38. T. F. Tuan, *Topophilia: A Study of Environmental Perception, Attitudes, and Values* (Englewood Cliffs, NJ: Prentice-Hall, 1974).

39. Klein, *This Changes Everything.*

40. "Native American Proverbs and Wisdom," Legends of America, last modified March 2017, http://www.legendsofamerica.com/na-proverbs2.html.

41. A. Schweitzer, *Out of My Life and Thought: An Autobiography*, trans. A. B. Lemke (Baltimore, MD: Johns Hopkins University Press, 2009).

42. Poynter, *The Human Experiment.*

Chapter 16

1. A. Alling et al., "Experiments on the Closed Ecological System in the Biosphere 2 Test Module," in *Ecological Microcosms*, eds. R. J. Beyers and H. T. Odum, (New York: Springer-Verlag, 1993), 463–469.

2. C. Hodges and R. Frye, "Soil Bed Reactor Work of the Environmental Research Lab of the University of Arizona in Support of the Biosphere 2 Project," in: *Biological Life Support Systems—Commercial Opportunities,* eds. M. Nelson and G. A. Soffen (Tucson, AZ: Synergetic Press, 1990), 33–40.

3. J. P. Allen, *Biosphere 2: The Human Experiment* (New York: Penguin, 1991).

4. K. Kelly, *Out of Control: The New Biology of Machines, Social Systems and the Economic World* (New York: Addison Wesley Publishing Company, 1994), 154.

5. R. L. Walford et al., "Physiologic Changes in Humans Subjected to Severe, Selective Calorie Restriction for Two Years in Biosphere 2: Health, Aging, and Toxicological Perspectives," *Toxicological Sciences* 52, no. 1 (1999): 61–65.

6. Kelly, *Out of Control,* 164–165.

7. M. Nelson, *The Emerging Lifestyle of the Biospherian*, unpublished manuscript, 1993.

8. D. Pauly, "Homo Sapiens: Cancer or Parasite?" *Ethics in Science and Environmental Politics* 14 (2014): 7–10, doi: 10.3354/esep00152.

9. F. Dyson, *Disturbing the Universe* (New York: Harper & Row, 1979).

10. P. Krugman, "This Can't Go On," NYTimes.com, November 4, 2003, http://www.nytimes.com/2003/11/04/opinion/this-can-t-go-on.html?mcubz=0.

11. M. Decoust, "Odyssey in Two Biospheres," Biosphere Foundation, http://biospherefoundation.org/project/odyssey-of-2-biospheres/.

12. W. S. Burroughs, "Four Horsemen of the Apocalypse," lecture at Planet Earth conference of the Institute of Ecotechnics, 1980, in *Man, Earth and the Challenges* (Santa Fe, NM: Synergetic Press, 1981), 153–168.

13. M. Nelson and A. Alling, "Biosphere 2 and Its Lessons for Long-Duration Space Habitats," *Space Manufacturing 9: The High Frontier Accession, Development and Utilization* (1993): 280–287.

Chapter 18

1. "Academic Politics Are So Vicious because the Stakes Are So Small," Quote Investigator, last modified August 22, 2013, http://quoteinvestigator.com/2013/08/18/acad-politics/.

2. M. Nelson, *The Wastewater Gardener: Preserving the Planet One Flush at a Time* (Santa Fe, NM: Synergetic Press, 2014); M. Nelson et al., "Worldwide Applications of Wastewater Gardens and Ecoscaping: Decentralised Systems which Transform Sewage from Problem to Productive, Sustainable Resource," in *Water and Wastewater Systems*, eds. K. Mathew, S. Dallas, and G. Ho (London: IWA Publications, 2008), 63–73.

3. J. Allen, D. Snyder, and K. Gray relating a conversation with R. Walford, personal communication, 1998.

4. L. Poppick, "Meet the Couple Who Could Be the First Humans to Travel to Mars," Wired.com, July 10, 2014, https://www.wired.com/2014/07/paragon-profile/.

5. "A Biospherian's Work Is Never Done," Tucson Local Media, June 29, 2005, http://www.tucsonlocalmedia.com/import/article_ed9bf835-f7c4-5d5c-a1b1-abe4e0a42cf2.html.

6. "Current Conservation and Research Projects," Biosphere Foundation, accessed June 10, 2017, http://biospherefoundation.org/current-conservation-projects/.

7. Ibid.

8. R. Reider, *Dreaming the Biosphere* (Albuquerque: University of New Mexico Press, 2009).

9. "Special Issue on Fourth International Symposium on Closed Ecological Systems: Biospherics and Life Support," *Life Support Biosphere Science* 4, no. 3–4 (1997): 87–181.

10. E. P. Odum, letter to *Science*, May 14, 1993, 878–879.

11. J. P. Cohn, "Biosphere 2: Turning an Experiment into a Research Station," *BioScience* 52, no. 3 (2002): 218–223.

12. B. D. V. Marino and H. T. Odum, "Biosphere 2: Introduction and Research Progress," *Ecological Engineering* 13 (1999): 3–14.

13. K. Arenson, "On Campus, a Gift Volvo Spotlights Questions about the Company's Involvement in Columbia's Environmental Education Programs," NYTimes.com, April 21, 1999, http://www.nytimes.com/1999/04/21/nyregion/campus-gift-volvo-spotlights-questions-about-company-s-involvement-columbia-s.html.

14. W. Broecker, "The Biosphere and Me," *GSA Today* 7, no. 6 (July 1996), http://www.geosociety.org/gsatoday/archive/6/7/pdf/i1052-5173-6-7-sci.pdf.

15. Reider, *Dreaming the Biosphere*.

16. "Second Batch of Volunteers Enter China's 'Lunar Palace,'" China Daily, July 10, 2017, http://www.chinadaily.com.cn/china/2017-07/10/content_30058251.htm; L. David, "China's 'Lunar Palace' for Space Research Tested on Earth," Space.com, June 16, 2014, http://www.space.com/26267-china-lunar-palace-space-research-mission.html.

17. H. T. Odum and B. V. D. Marino, eds., "Biosphere 2 Research Past and Present," reprinted from *Ecological Engineering Special Issue* 13, no. 1–4 (1999).

18. Reider, *Dreaming the Biosphere*.

19. H. T. Odum, "Scales of Ecological Engineering," *Ecological Engineering* 6 (1996): 7–19.

20. J. Lovelock, *Gaia: A New Look at Life on Earth* (Oxford: Oxford University Press, 1979).

21. J. Allen, "Biospheric Theory and Report on Overall Biosphere 2, Design and Performance During Mission One (1991–1993)," *Life Support and Biosphere Science* 4, no. 3–4 (1997): 95–108.

22. R. L. Walford, "Biosphere 2 as Voyage of Discovery: The Serendipity from Inside," *Bioscience* 52, no. 3 (2002): 259–263.

23. H. D. Adams et al., "Temperature Sensitivity of Drought-Induced Tree Mortality Portends Increased Regional Die-Off Under Global-Change-Type Drought, *Proceedings of the National Academy of Sciences* 106, no. 17 (2009): 7,063–7,066.

24. Reider, *Dreaming the Biosphere*.

25. Odum, "Scales of Ecological Engineering."

26. D. S. Ellsworth et al., "Photosynthesis, Carboxylation and Leaf Nitrogen Responses of 16 Species to Elevated pCO_2 Across Four Free-Air CO_2

Enrichment Experiments in Forest, Grassland and Desert," *Global Change Biology* 10 (2004): 2,121–2,138, doi: 10.1111/j.1365–2486.2004.00867.x.

27. Cohn, "Biosphere 2: Turning an Experiment into a Research Station."

28. K. L. Griffin et al., "Leaf Respiration Is Differentially Affected by Leaf vs. Stand-Level Nighttime Warming," *Global Change Biology* 8, no. 5 (2002): 479–485.

29. C. Langdon et al., "Effect of Calcium Carbonate Saturation State on the Calcification Rate of an Experimental Coral Reef," *Global Biogeochemical Cycles* 14, no. 2 (2000): 639–654; C. Langdon et al., "Effect of Elevated CO_2 on the Community Metabolism of an Experimental Coral Reef," *Global Biogeochemical Cycles* 17, no. 1 (2003): 1,011, doi:1010.1029/2002GB001941.

30. W. C. Harris and L. J. Graumlich, "Biosphere 2: Sustainable Research for a Sustainable Planet," Columbia University, accessed June 12, 2017, http://www.columbia.edu/cu/21stC/issue-4.1/harris.html.

31. G. Lin et al., "Ecosystem Carbon Exchange in Two Terrestrial Ecosystem Mesocosms under Changing Atmospheric CO_2 Concentrations," *Oecologia* 119 (1999): 97–108.

32. G. Lin, et al., "An Experimental and Model Study of the Responses in Ecosystem Exchanges to Increasing CO_2 Concentrations Using a Tropical Rainforest Mesocosm," *Australian Journal of Plant Physiology* 25 (1998): 547–556; G. Lin et al., "Sensitivity of Photosynthesis and Carbon Sinks in World Tropical Rainforests to Projected Atmospheric CO_2 and Associated Climate Changes," *Proceedings 12th International Congress on Photosynthesis* (2001).

33. F. A. Bernstein, "Sprawl Outruns Arizona's Biosphere," NYTimes.com, May 28, 2006, http://www.nytimes.com/2006/05/28/realestate/28nation.html?oref=slogin.

34. "Landscape Evolution Observatory," Biosphere 2, accessed June 15, 2017, http://biosphere2.org/research/projects/landscape-evolution-observatory.

35. "Rainforest Drought," Biosphere 2, accessed June 15, 2017, http://biosphere2.org/research/projects/rainforest-drought.

36. "Desert Sea," Biosphere 2, accessed June 15, 2017, http://biosphere2.org/research/projects/desert-sea.

37. "Rockubators," Biosphere 2, accessed June 16, 2017, http://biosphere2.org/research/projects/rockubators; "Model City," Biosphere 2, accessed June 17, 2017, http://biosphere2.org/research/model-system/model-city.

38. "Trace Gas Laboratory," Biosphere 2, accessed June 16, 2017, http://biosphere2.org/research/laboratory/trace-gas-laboratory; "Critical Zone," Biosphere 2, accessed June 12, 2017, http://biosphere2.org/research/themes/critical-zone.

Conclusion

1. T. S. Kuhn, *The Structure of Scientific Revolutions* (Chicago: University of Chicago Press, 1970).

2. American Heritage® Dictionary of the English Language, 5th ed. (New York: Houghton Mifflin Harcourt, 2011), http://www.thefreedictionary.com /paradigm.

3. "International Press Release to Support Citizens in the Fight Against Fracking in In Salah (Southern Algeria)," Friends of the Earth, March 27, 2015, http:// www.amisdelaterre.org/Communique-international-de.html.

4. E. Barbier, "Account for Depreciation of Natural Capital," *Nature* 515 (November 2014): 32–33.

5. E. O. Wilson, *Half-Earth: Our Planet's Fight for Life* (New York: Liverlight, 2016).

6. "Max Planck," WikiQuote, last modified August 5, 2017, http://en.wikiquote .org/wiki/Max_Planck.

7. S. Springer, K. Birch, and J. MacLeavy, eds., *The Handbook of Neoliberalism* (New York: Routledge, 2016).

8. World Commission on the Environment and Development, *Our Common Future* (Oxford: Oxford University Press, 1987).

9. C. J. Jeroen and M. van den Bergh, "Ecological Economics: Themes, Approaches, and Differences with Environmental Economics," *Regional Environmental Change* 2, no. 1 (2001): 13–23; T. Juniper, *What Has Nature Ever Done for Us: How Money Really Grows on Trees* (Santa Fe, NM: Synergetic Press, 2013); P. Hawken, A. Lovins, and H. Lovins, *Natural Capital: Creating the Next Industrial Revolution* (New York: Simon and Schuster, 1999).

10. OECD Better Life Index, http://www.oecdbetterlifeindex.org/.

11. *UnU-iHDP and UNEP Inclusive Wealth Report 2012: Measuring Progress toward Sustainability*, (Cambridge, MA: Cambridge University Press, 2012).

12. "Bhutan's Gross National Happiness Index," Oxford Poverty and Human Development Initiative, accessed September 13, 2017, http://www.ophi.org.uk/policy /national-policy/gross-national-happiness-index/.

13. D. H. Meadows, *The Limits to Growth: A Report for the Club of Rome's Project on the Predicament of Mankind* (New York: Universe Books, 1972).

14. D. H. Meadows, D. L. Meadows, and J. Randers, *Beyond the Limits: Confronting Global Collapse, Envisioning a Sustainable Future* (Post, VT: Chelsea Green Books, 1992).

15. R. Costanza et al., "Changes in the Global Value of Ecosystem Services," *Global Environmental Change* 26 (2014): 152–158.

16. P. H. Gleick, "Water, Drought, Climate Change, and Conflict in Syria," *Weather Climate and Society* 6 (2014): 331–340, doi: http://dx.doi.org/10.1175/WCAS-D-13-00059.1.

17. N. Meyers, "Environmental Refugees: A Growing Phenomenon of the 21st Century," *Royal Society*, October 19, 2001, http://www.ncbi.nlm.nih.gov/pmc/articles/PMC1692964/pdf/12028796.pdf.

18. J. Vidal, "Global Warming Could Create 150 Million 'Climate Refugees' by 2050," *The Guardian*, November 2, 2009, https://www.theguardian.com/environment/2009/nov/03/global-warming-climate-refugees.

19. Hawken, Lovins, and Lovins, *Natural Capital*; E. U. von Weizsacker et al., *Factor Five: Transforming the Global Economy Through 80% Improvements in Resource Productivity: A Report to the Club of Rome*, Earthscan, 2009.

20. "Industrial Ecology," accessed September 3, 2017, http://www.gdrc.org/sustdev/concepts/16-l-eco.html.

21. UNCTAD, *Trade and Environment Review 2013, Wake Up Before It's Too Late: Make Agriculture Truly Sustainable Now for Food Security in a Changing Climate*, 2013, http://unctad.org/en/PublicationsLibrary/ditcted2012d3_en.pdf.

22. T. Ghose, "The Secret to Curbing Population Growth," April 29, 2013, http://www.livescience.com/29131-economics-drives-birth-rate-declines.html.

23. Research Institute of Organic Agriculture, *The World of Organic Agriculture, Statistics and Emerging Trends 2014*, FiBL and International Federation of Organic Agricultural Movements, https://www.fibl.org/fileadmin/documents/shop/1636-organic-world-2014.pdf.

24. R. B. Fuller, *Operating Manual for Spaceship Earth* (New York: Simon and Schuster, 1969).

25. ClimateDebt.org, accessed September 3, 2017, http://www.climate-debt.org.

26. "Solar Electricity Costs," accessed September 4, 2017, http://solarcellcentral.com/cost_page.html.

27. F. Guerrini, "Solar Power to Become Cheapest Source of Energy in Many Regions by 2025, German Experts Say," March 31, 2015, http://www.forbes.com/sites/federicoguerrini/2015/03/31/solar-power-to-become-cheapest-source-of-energy-in-many-regions-by-2025-german-experts-say/.

28. C. Chait, "The Year Humans Finally Got Serious about Saving Themselves from Themselves," *New York* magazine, September 7, 2015, http://nymag.com/daily/intelligencer/2015/09/sunniest-climate-change-story-ever-read.html.

29. S. Anup, "World Military Spending," Global Issues, 2013, http://www.globalissues.org/article/75/world-military-spending.

30. *Stern Review: The Economics of Climate Change*, http://mudancasclimaticas.cptec.inpe.br/~rmclima/pdfs/destaques/sternreview_report_complete.pdf.

31. N. Klein, *This Changes Everything: Capitalism vs. the Climate* (New York: Simon and Shuster, 2014).

32. R. Louv, *Last Child in the Woods* (Chapel Hill, NC: Algonquin Books, 2005).

33. United Nations Environment Program, "Cities as Sustainable Ecosystems," http://www.unep.or.jp/ietc/Publications/Freshwater/FMS7/9.asp.

34. R. A. Clay, "Green is Good for You, Monitor on Psychology," *American Psychological Association* 32, no. 4 (April 2001): 40.

35. L. Górska-Kłęk, K. Adamczyk, and S. Krzysztof, "Hortitherapy—Complementary Method in Physiotherapy, *Fizjoterapia* 17, no. 4 (January 2009).

36. W. S. Burroughs, "William S. Burroughs on Censorship," *Transatlantic Review* 11 (winter 1962), https://beatpatrol.wordpress.com/2008/11/23/william-s-burroughs-on-censorship-1962/.

37. "The History, Philosophy and Practice of Buddhism," accessed September 2, 2017, http://www.buddha101.com/p_path.htm.

38. S. J. Gould, *Eight Little Piggies* (New York: W. W. Norton, 1993).

39. J. Wilson, *Mindful America: Meditation and the Mutual Transformation of Buddhism and American Culture* (Oxford: Oxford University Press, 2014).

40. "Curricular Modules of Study," California Institute of Integral Studies, accessed September 3, 2017, https://www.ciis.edu/public-programs/certificate-programs/certificate-in-psychedelic-assisted-therapies-and-research/curricular-modules-of-study.

41. D. E. Sherkat and C. G. Ellison, "Structuring the Religion-Environment Connection: Identifying Religious Influences on Environmental Concern and Activism," *Journal for the Scientific Study of Religion* 46 (2007): 71–85.

42. Apostolic Exhortation, *Evangelii Gaudium*, November 24, 2013, 56: AAS 105, 1043.

43. C. Hodges, "Biosphere 2: The Key Variables," presentation at Biospheres Conference 2, Oracle, AZ, September 1985, quoted in Reider, *Dreaming the Biosphere*.

44. "Transcript: Rep. John Lewis's Speech on 50th Anniversary of the March on Washington," *The Washington Post*, August 28, 2013, https://www.washingtonpost.com/politics/transcript-rep-john-lewiss-speech-on-50th-anniversary-of-the-march-on-washington/2013/08/28/fc2d538a-100d-11e3-8cdd-bcdc09410972_story.html?utm_term=.0b6fff33b13a.

Index

Page numbers in **bold** indicate illustrations.

About the Author

Dr. Mark Nelson is a founding director of the Institute of Ecotechnics and has worked for several decades in closed ecological system research, ecological engineering, the restoration of damaged ecosystems, desert agriculture and orchardry, and wastewater recycling. He is chairman of the Institute of Ecotechnics, vice chairman of Global Ecotechnics Corporation, and head of Wastewater Gardens International. Nelson received his PhD in environmental engineering sciences from the University of Florida in 1998 and an MS from the University of Arizona in 1995. He was a member of the eight-person biospherian crew for the first two-year closure experiment at Biosphere 2 from 1991 to 1993. His books include *The Wastewater Gardener: Preserving the Planet One Flush at a Time* (Synergetic Press, 2014), and he co-authored *Life Under Glass: The Inside Story of Biosphere 2* (Synergetic Press, 1993). His research papers have been published in *Life Support and Biosphere Science*, *BioScience*, *Ecological Engineering*, and *Advances in Space Research*. Nelson was a contributing editor for the journal *Life Support and Biosphere Science* from 1993 to 2002, *Advances in Space Research* from 2000 to 2013, and he is currently an associate editor of *Life Sciences in Space Research*. Nelson was awarded the Yuri Gagarin Jubilee Medal in 1993 by the Russian Cosmonautics Federation and was elected a Fellow of the Explorers Club (1994) and a Fellow of the Royal Geographical Society (2001).